SERBER SAYS:
About Nuclear Physics

ROBERT SERBER

World Scientific Lecture Notes in Physics Vol. 10

ſERBER ſAYſ:
About Nuclear Physicſ

Robert Serber

Columbia University, USA

World Scientific

Published by

World Scientific Publishing Co. Pte. Ltd.

5 Toh Tuck Link, Singapore 596224

USA office: 27 Warren Street, Suite 401-402, Hackensack, NJ 07601

UK office: 57 Shelton Street, Covent Garden, London WC2H 9HE

British Library Cataloguing-in-Publication Data
A catalogue record for this book is available from the British Library.

SERBER SAYS
About Nuclear Physics

ISSN 0218-026X

ISBN-13 978-9971-5-0158-7
ISBN-10 9971-5-0158-9
ISBN-13 978-9971-5-0376-5 (pbk)
ISBN-10 9971-5-0376-X (pbk)

TO
MY TEACHERS
JOHN HASBROUCK VAN VLECK
AND
J. ROBERT OPPENHEIMER

CONTENTS

SERBER SAYS:
About Nuclear Physics

CHAPTER 1

SOME ARGUMENTS
CONCERNING NUCLEAR FORCES

The properties of nuclei in their ground states
resemble, in important respects, those of ordinary solids
and liquids. The binding energy is approximately
proportional to the number of nucleons, A, and the mean
density is is nearly independent of A. The curve of binding
energy per particle versus A, shown in Fig.1.1, is similar
to that of a charged liquid drop. If the liquid has a
particle density ρ, an energy density $-\epsilon\rho$, a surface tension
σ and a charge Ze, the energy per particle would be

$$E/A = -\epsilon + \sigma S/A + (3/5) Z^2 e^2 / RA, \qquad (1.01)$$

where S is the surface area and R is the radius. The first
term is the volume energy, the second the surface energy
and the third the Coulomb energy.

The decrease in $-E/A$ for small A is interpreted as due
to the repulsive surface term which, since

$$A = (4\pi/3)\rho R^3, \quad S = 4\pi R^2, \qquad (1.02)$$

is proportional to $1/A^{1/3}$. The decrease for large A is due

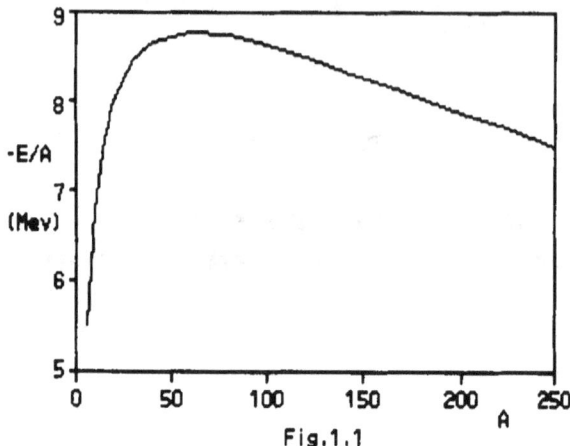

Fig.1.1

to the repulsive Coulomb term which, since Z/A decreases with increasing A, goes a little less rapidly than $A^{2/3}$. Using observed nuclear radii we can easily check that the decrease in −E/A of 1.2 Mev between A=60 and A=240 is in the right ball-park. The first of the relations (1.02) can be written

$$R = r_0 A^{1/3}.$$

We take $r_0=1.4$ f, which has the convenient property that it is just half of the classical electron radius, $r_0 = (1/2)e^2/mc^2$. The Coulomb energy becomes $(3/5)(Z^2/A^{4/3}) \cdot 2mc^2 = (3/5)(Z^2/A^{4/3})$ Mev. This gives 1.9 Mev for $^{26}Fe^{56}$, 3.4 Mev for $^{92}U^{238}$, a difference of 1.5 Mev.

One might infer from this that the forces between nuclei are similar to those between atoms (Fig.1.2), an inference that is indeed true. However in simpler minded days (1932) Heisenberg[1] argued that a potential such as that of Fig.1.2 would only arise in the interaction of complex structures, and nucleons, being "elementary particles", should show more "elementary" behavior. A

---

[1] W.Heisenberg, Z. Physik 77,1 (1932).

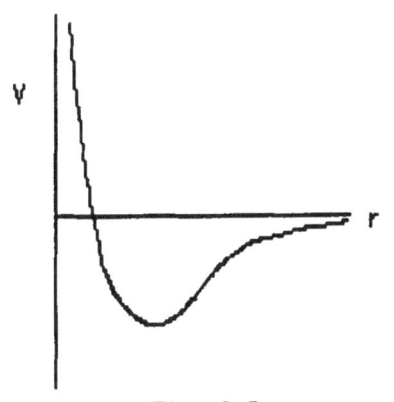

Fig. 1.2

Coulomb-like interaction is an elementary behavior but it would not lead to the required properties. In the liquid (nuclear matter) the energy density is locally determined and does not depend on the size of the system. The potential energy of a particle in the liquid is $\int V(r)\rho d\underline{r}$. If $V(r)\approx 1/r^n$ for large r, the contribution from distant parts of the liquid is $4\pi\rho\int(r^2/r^n)dr$. Thus n must be greater than 3 for the distant parts not to contribute. Since a $1/r^n$ potential with n>3 is too singular at the origin, we conclude that the force has a range.

A simple generalization of the Coulomb interaction, $V=e\bar{\phi}$,

$$\bar{\phi}=e/4\pi r, \qquad (1.03)$$

to one which has a finite range is the Yukawa interaction, $V=-g\bar{\phi}$,

$$\bar{\phi}=ge^{-\mu r}/4\pi r. \qquad (1.04)$$

A minus sign has been inserted to make the force attractive. For an extended source the $\bar{\phi}$ of (1.03) satisfies Laplace's equation,

$$\Delta\bar{\phi}=-e\rho. \qquad (1.05)$$

The $\bar{\phi}$ of (1.04) can be regarded as elementary in that it also satisfies a simple differential equation,

$$\Delta\bar{\phi}-\mu^2\bar{\phi}=-g\rho. \qquad (1.06)$$

As Yukawa[2] pointed out, the generalization of (1.06) to a time dependent relativistic equation replaces the Δ by the

---

[2]H. Yukawa, Proc. Phys. Math. Soc., Japan 17, 48 (1935).

$\square$, and the free-field equation,

$$\square\xi-\mu^2\xi=0,$$

has plane wave solutions, $\xi=\exp[i(\mathbf{p}\cdot\mathbf{r}-Et)]$, with $E^2=\mu^2+p^2$, the right relation for a particle of mass μ. (Here and hereafter we take $h/2\pi=c=1$.) In the lingo of field theory, ξ is a neutral scalar meson field. The range of the nuclear force is thus related to the Compton wavelength of the particle field carrying the interaction.

However, as Heisenberg pointed out, an interaction such as (1.04) does not lead to the proper nuclear properties. In the interior of a nucleus, where ρ is constant, the potential energy of a particle is

$$V=\int V(r)\rho d\mathbf{r}=\rho V_0, \text{ with } V_0=\int V(r)d\mathbf{r}, \tag{1.07}$$

and thus increases linearly with ρ. In order to have equilibrium at a finite ρ, the kinetic energy must produce enough pressure to balance the attractive forces. Let us estimate the kinetic energy by treating the nuclear matter as a degenerate Fermi gas.

In the Fermi gas each state occupies a volume h^3 in phase space, or in our units, with $h/2\pi=1$, a volume $(2\pi)^3$. If there are A particles in a volume V_{ol}, and the momentum space is filled up to the Fermi momentum p_f, we thus have $A=4(4\pi/3)p_f^3 V_{ol}/(2\pi)^3$. The factor 4 arises because four particles can be put in each state, protons and neutrons with spins up or down (for simplicity, take the number of protons and neutrons to be the same). We then have

$$\rho=A/V_{ol}=(2/3\pi^2)p_f^3. \tag{1.08}$$

The mean kinetic energy of a nucleon is

$$T=\int_0^{p_f}(p^2/2M)d\mathbf{p}/\int_0^{p_f}d\mathbf{p}=(1/2M)\int_0^{p_f}p^4dp/\int_0^{p_f}p^2dp=(3/5)T_f, \tag{1.09}$$

with $T_f = p_f^2/2M$. The mean energy per nucleon (neglecting surface and Coulomb energy) is

$$E/A = T + \tfrac{1}{2}V = (3/5)(1/2M)(3\pi^2/2)^{2/3}\rho^{2/3} + \tfrac{1}{2}V_0\rho. \qquad (1.10)$$

The mean potential energy per nucleon is $(1/2)V$, rather than V, since V would include the interaction of particle A with B plus the interaction of B with A.[3] Equation (1.10) is plotted in Fig.3a; it leads to no stable configuration.

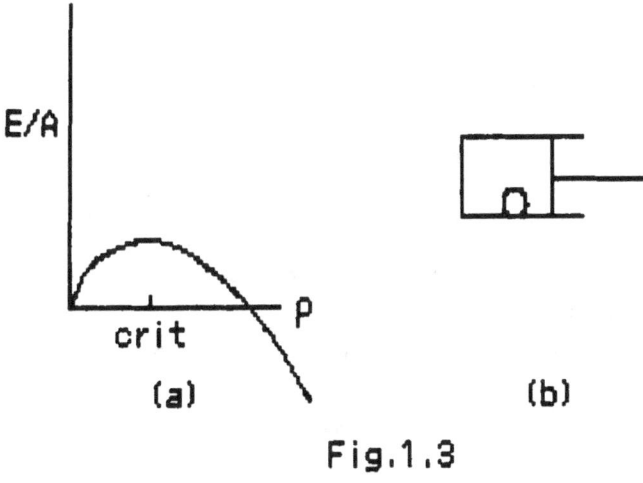

Fig.1.3

---

[3] The energy density is
$$\epsilon = \rho(E/A) = (3/5)(1/2M)(3\pi^2/2)^{2/3}\rho^{5/3} + \tfrac{1}{2}V_0\rho^2.$$ The energy of a nucleon at the top of the Fermi sea (the binding energy of the last nucleon) is, since $E = \epsilon V_{0l}$, $A = \rho V_{0l}$, $dE/dA = d\epsilon/d\rho = T_f + V$. Consistency requires that at equilibrium this equal (1.10). Setting the two expressions equal leads to the virial relation, $T_f = -(5/4)V$, which shows the nucleus is unbound. The reader can verify that at the maximum in Fig.1.3(a) the virial relation is satisfied.

The physical situation is illustrated by Fig.3b.
Imagine the Fermi gas in a cylinder being compressed by a
piston.   At low density the gas exerts a positive pressure
on the piston. At a critical density we reach a point of
unstable equilibrium, the attractive forces between
nucleons just balancing the Fermi pressure, and the
pressure on the piston is zero. If the piston is pushed a
little further the system collapses to a droplet on the
cylinder floor. In our approximation the density would be
infinite; actually the nucleus would collapse to a radius
near the range of forces so (1.07) would fail, or p_f would
become large enough so that relativistic corrections became
important.   Lee and Wick[*] have pointed out that if
relativistic formulas are used for the kinetic and
potential energies there is a solution of the problem, but
one which gives a binding energy much larger than that of
ordinary nuclei.

This difficulty is avoided by Heisenberg's supposition
that the nuclear forces are exchange forces between protons
and neutrons. By this is meant that if a proton is an a
state (including spin) $u_1(r)$ and the neutron in a state
$u_2(r)$ the interaction energy, rather than being the
ordinary term

$$\int u_1(r_P)^* u_2(r_N)^* V(r_P - r_N) u_1(r_P) u_2(r_N) dr_P dr_N, \qquad (1.11)$$

is instead an exchange term,

$$\int u_1(r_N)^* u_2(r_P)^* V(r_P - r_N) u_1(r_P) u_2(r_N) dr_P dr_N, \qquad (1.12)$$

that is, the neutron and proton coordinates and spins are
interchanged in the final state. Exchange terms were
familiar in atomic and molecular theory, and Heisenberg's
justification for the exchange force was based on an
analogy with the H_2^+ ion. He suggested a model in which the

---

T.D. Lee and G.C. Wick, Phys. Rev. D $\underline{9}$, 2291 (1974).

neutron was regarded as proton plus electron, the exchange
force arising from exchange of the electron between the two
protons. A more plausible interpretation was supplied by
Yukawa. The interaction (1.04) can be viewed as due to the
exchange of an uncharged meson, as illustrated in Fig.1.4(a).

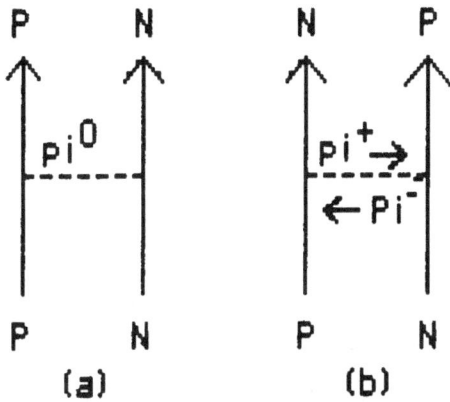

Fig.1.4

If the mesons were charged we would have Fig.1.4(b), which
exchanges the proton and neutron in the final state. We
then have, in place of (1.04),

$$V_{NP}=\pm(g^2/4\pi)(e^{-\mu r}/r)P^H, \tag{1.13}$$

where P^H is defined to be an operator which exchanges the
coordinates and spins of neutron and proton. Since Yukawa
envisaged only charged mesons there would be no N-N or P-P
forces of order g^2 (single meson exchange), in accord with
the Heisenberg model. The $\pm$ sign in (1.13) depends on the
transformation properties supposed for the meson field, $\underline{\delta}$.
Yukawa originally supposed $\underline{\delta}$ was the fourth component of a
four-vector field, in analogy to the electrostatic
potential, in which case the + sign would apply. For a
scalar $\underline{\delta}$ the minus sign is correct. The latter is required
for the N-P interaction, as can most easily be seen by
considering the deuteron: the potential must be attractive

for the bound state, which is S^3 and thus has $P^H=1$.

If we write V_q for the Fourier component of $V(r)$, i.e.

$$V(\underline{r})=\int V_q e^{i\underline{q}\cdot\underline{r}}d\underline{q}/(2\pi)^3, \qquad V_q=\int V(\underline{r})e^{-i\underline{q}\cdot\underline{r}}d\underline{r},$$

one sees that if u_1 and u_2 are plane waves of momentum $\underline{p}$ and $\underline{p}'$, (1.11) gives V_0, while (1.12) gives V_q, with $\underline{q}=\underline{p}-\underline{p}'$. Note that the V_0 defined here is the same as that of (1.07).

As an example of the difference an exchange force makes, let us consider the Yukawa potentials, (1.04) or (1.13). Here

$$V_q=-g^2/(\mu^2+q^2), \qquad\qquad (1.14)$$

and $V_0=-g^2/\mu^2$ (note that, for $\rho=const.$, the solution of (1.06) is $\bar{g}=(g/\mu^2)\rho$ which gives $V=-(g^2/\mu^2)\rho$, in agreement with (1.07)). In the high density limit, $p_f \gg \mu$, the exchange term (1.14) is smaller than V_0 by two powers of q. Thus in (1.10) the potential energy term will be proportional to p_f rather than to $\rho \approx p_f^3$. With a repulsive kinetic energy proportional to p_f^2 and an attractive potential energy proportional to p_f, (1.10) will have a minimum, at negative E/A, for a finite p_f.

Heisenberg's exchange force, while ingenious, is not the true explanation of the "saturation of nuclear forces". It is now known from high energy scattering experiments that the nuclear forces are repulsive at small distances, less than 0.5 f, like Fig.1.2. The effect of the "repulsive core" can be estimated by using the results of the theory of a hard-sphere Fermi gas in the form given by Bohr and Mottleson[*]. The kinetic energy (1.09) is replaced by

---

[*] A. Bohr and B.R. Mottleson, Nuclear Structure (Benjamin, New York, 1969), Vol.1, p. 256.

$$T = \frac{3}{5} \frac{p_f^2}{(2M)} \frac{1}{[1-(5/3\pi)r_c p_f]^2}, \tag{1.15}$$

where r_c is the core radius. The form of (1.15) is a little surprising, since one might expect to estimate the correction by reducing V_{ol} in (1.08) by the excluded volume, $(4\pi/3)r_c^3 A$. This argument would lead to a correction factor $1/[1-(4\pi/3)r_c^3 \rho]^{2/3}$, i.e. a correction proportional to $(r_c p_f)^3$ rather than one linear in $r_c p_f$.

With the nuclear radius written $R=r_0 A^{1/3}$ we have

$$\rho = 3/(4\pi r_0^3), \quad p_f = (9/8\pi)^{1/3}/r_0, \tag{1.16}$$

the second relation following from (1.08). The excluded volume argument gives T a correction factor $1/[1-(r_c/r_0)^3]^{2/3}$. The value of r_0 determined from high energy electron scattering experiments is $r_0=1.1$ f. For $r_c=0.5$ f the correction would be only 7%. This lead to a long-held belief that a core radius of this size was too small to stabilize the nucleus at so low a density, and thus an alternative such as Heisenberg's was necessary. On the other hand, (1.15) gives a factor $1/[1-(5/3\pi)(9\pi/8)^{1/3}(r_c/r_0)]^2 = 1/[1-0.8r_c/r_0]^2 = 2.5$, a comparatively enormous effect.

Some feeling for the reason $r_c p_f$ appears linearly in (1.15) may perhaps be found in the following argument. In terms of the variable r_0, rather than p_f, (1.15) is of the form

$$T = B/(r_0 - 0.8r_c)^2,$$

with B a constant. Consider a particle inside a sphere of of radius r_0, with a wave function $\phi(r)=u(r)/r$ satisfying some simple boundary condition at the surface, e.g. $u'(r_0)=0$. The solution is $u=\sin(pr)$, $pr_0=\frac{1}{2}\pi$, and the energy is

$$T = p^2/2M = (1/2M)(\pi/2)^2/r_0^2.$$

If a repulsive core is added at the center (or an excluded
shell at the edge) we would have $p(r_0-r_c)=\frac{1}{2}\pi$, and an energy

$$T=(1/2M)(\pi/2)^2/(r_0-r_c)^2.$$

To continue the argument, write the binding energy in
terms of r_0; it then takes the form (see (1.10)

$$E/A=T+\tfrac{1}{2}V=B/(r_0-a)^2-C/r_0^3,$$

where we have written $a=0.8r_c$. The condition for a minimum
is

$$\frac{d}{dr_0}(\frac{E}{A})=-\frac{2T}{r_0-a}-\frac{3V}{2r_0}=0,$$

or

$$1-(a/r_0)=(2/3)T/(-\tfrac{1}{2}V).$$

If we suppose that in nuclear matter the binding energy is
a small difference between large kinetic and potential
energies, $T/(-\tfrac{1}{2}V)\approx1$, and

$$1-a/r_0=2/3,\ a/r_0=1/3,\ r=3a=3\times0.8r_c.$$

For $r_c=0.5$ f this gives $r_0=1.2$ f, close to the observed
$r_0=1.1$ f. Thus a repulsive core of the observed range can
account for the actual nuclear density.

For $r_0=1.1$ f, the Fermi energy is $T_f=42$ Mev and T,
given by (1.15), is $T=(3/5)\times2.5\times42=63$ Mev. The empirical
volume binding energy, the constant ϵ in (1.01), is 15 Mev.
Thus $(1/2)V=-78$ Mev. These numbers are not to be taken too
literally, but they give an idea of the magnitudes
involved.

It should also be observed that these arguments depend
only on the the volume integral of the potential, e.g. for
a Yukawa potential on g^2/μ^2, and do not separately
determine the strength of the interaction or its range.

A more complete but equally simple-minded model,
discussing all three terms in (1.01), will be found in
Appendix A, "A Simple Nuclear Model".

CHAPTER 2

THE NEUTRON—PROTON FORCE

The most direct evidence on the properties of the nuclear forces would be expected to come from the study of the two-nucleon system, primarily from nucleon-nucleon scattering experiments. In the early days of nuclear physics, all through the thirties and well into the forties, the evidence from this source was limited. To determine the shape of the potential one requires a probe whose wavelength is shorter than the size to be distinguished. In the case of nucleon-nucleon scattering by a potential of range R, this requires at least

$$1/p < R, \qquad\qquad (2.01)$$

or, since the kinetic energy in the center of mass system is $E = p^2/2\mu$ with μ the reduced mass, equal to half the nucleon mass, $\sqrt{(1/ME)} < R$, $E > 1/(MR^2)$. For a reasonable choice of R, R=1.2 f, this gives

$$E > M/(MR)^2 = 939 \times (0.21/1.2)^2 = 28.8 \text{ Mev},$$

where we have used $1/M = 0.21$ f, $M = 939$ Mev. The energy of the bombarding nucleon in the laboratory system is twice the center of mass energy, so we require $E_{Lab} > 58$ Mev.

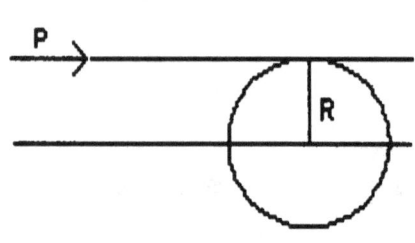

Fig.2.1

Another form of the argument is illustrated by Fig.2.1. The angular momentum of the particle passing at distance R, and thus the maximum angular momentum which will be scattered, is l=pR. For l=1 (p wave scattering) we must have pR>1, the same condition as (2.01). Thus for the particle energies available in the thirties, below 10 Mev, we can expect only s wave scattering. This can furnish important information on the forces, but far from a complete description.

For proton-proton scattering, even in the s state, classically one would require sufficient energy to surmount the Coulomb barrier before the region of nuclear forces is reached. One would need

$E > e^2/R = e^2/(mc^2 R) mc^2 = (2.8/1.2) \times 0.51 = 1.2$ Mev, $E_{Lab} > 2.4$ Mev.

Here m is the electron mass, and we have used $e^2/mc^2 = 2.8$ f, $mc^2 = 0.51$ Mev and taken R=1.2 f. Because of the quantum mechanical barrier penetration effect this is an over-estimate, but energies greater than 0.5 Mev are still required. Before the middle thirties proton beams of such energies were not available; it was not till 1935 that the first P-P and N-P scattering experiments were done, nearly simultaneously.

SEC.2I THE DEUTERON

For the N-P system we have an additional item of
information: the existence of a bound state, the deuteron,
with a binding energy of ϵ_D=2.22452 Mev. If we suppose the
N-P interaction is an attractive central force, the lowest
state would be expected to be $^3S_1$ or $^1S_0$. In fact, the
deuteron has J=1. Its magnetic moment is μ_D=0.857393
nuclear magnetons, while the sum of the proton and neutron
moments is 0.87954 (μ_P=2.79270, μ_N=-1.91316). This close
agreement indicates that any contribution of an orbital
angular momentum is small, and that, to a good
approximation, the deuteron can indeed be described as
being in an s state.

If we write for the bound state wavefunction ϕ=u(r)/r,
the s wave radial equation is

$$u'' + M(-\epsilon - V(r))u = 0. \qquad (2.02)$$

This equation will have a solution only if V(r) is
sufficiently large. In this respect, the radial problem,
$0 \leq r < \infty$, differs from the one dimensional problem, $-\infty < r < \infty$.
In the latter case any attractive potential produces a
bound state, as indicated in Fig.2.2(a). Note that outside

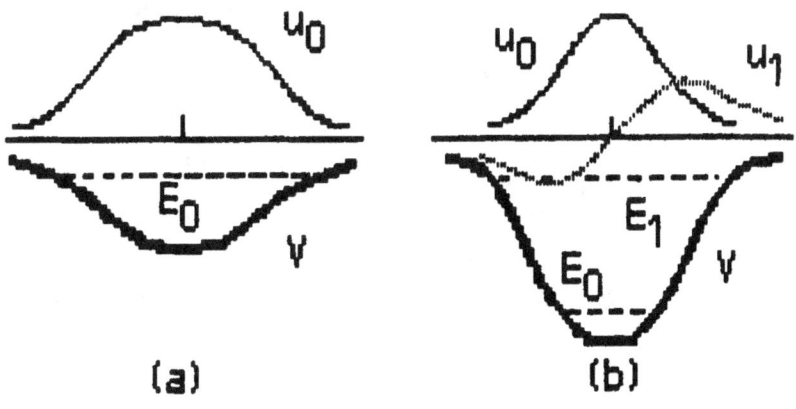

(a)   (b)

Fig.2.2

the range of the force, i.e. for r sufficiently large that
− V(r)<<∈, the solutions of (2.02) are

$$u \simeq \exp[\pm(M\epsilon)^{1/2}r].\hspace{2cm}(2.03)$$

Now suppose that the magnitude of V(r) is gradually
increased, keeping the same shape. At a certain critical
magnitude a second bound state appears with zero energy,
$\epsilon_1 = 0$; if the magnitude is further increased ϵ_1 increases
also. If V(r) is even in r, V(−r)=V(r), it is evident that
u_1 is odd, as shown in Fig.2.2b. For the radial problem,
on the half interval, the lowest state $u_0(r)$ is
unacceptable, since it fails to satisfy the boundary
condition u(0)=0 (remember ∅=u/r). The odd solution, $u_1(r)$,
although the first excited state of of the one dimensional
problem, is the lowest state of the radial problem.

  The critical potential, which just produces a bound
state with ∈=0, can easily be found for a square well
potential (whose only virtue is that it permits an easy
solution of the wave equation). For a square well of range
R, V(r)=−V_0 for r<R, V(r)=0 for r>R. The solution of
(2.02), with ∈=0 and satisfying the boundary condition at
the origin, is u=sin(pr), with p=√(MV_0). According to
(2.03), for ∈=0 u is a constant outside the well, which
thus must contain exactly a quarter wavelength, pR=½π, or

$$V_0 = \pi^2/(4MR^2).\hspace{2cm}(2.04)$$

  For a Yukawa potential, V=−$(g^2/4\pi)e^{-\mu r}/r$, the wave
function takes the form, if rescaled by writing y=μr,

$$d^2u/dy^2+[(g^2/4\pi)(M/\mu)(e^{-y}/y)-(M\epsilon/\mu^2)]u=0.\hspace{1cm}(2.05)$$

For ∈=0 only the combination $f^2=(g^2/4\pi)M/\mu$ appears. By
numerical integration we find the boundary condition,
u≃const as r→∞, is satisfied for f^2=1.68. Finite binding
requires a larger value of f^2. In the zero-range limit,
1/μ→0, we see from (2.05) that the change in f^2 is is of

order $M\epsilon/\mu^2$, which, for fixed ϵ, approaches zero. Table 2.1 shows, for three finite ranges, the value of f^2 required to give the observed binding energy of the deuteron, ϵ_D=2.2245 Mev.

For a Yukawa potential plus a repulsive core of radius 0.5 f the strength of coupling to give ϵ_D=2.2245 Mev is shown in Table 2.2.

Table 2.1					
$1/\mu$	f^2	$g^2/4\pi$	a_t	r_t (2.38)	r_t (2.35)
0	1.68	∞	4.318	0	0
0.7	2.04	0.613	4.964	1.13	1.11
1.05	2.22	0.444	5.211	1.48	1.44
1.4	2.39	0.359	5.409	1.74	1.65

Table 2.2					
$1/\mu$	f^2	$g^2/4\pi$	a_t	r_t (2.38)	r_t (2.35)
0	∞	∞	4.819	0.90	0.90
0.35	21.21	12.74	5.292	1.59	1.60
0.7	8.45	2.54	5.639	2.03	2.05
1.05	6.34	1.27	5.916	2.33	2.35
0.48	12.74	5.58	5.425	1.76	1.78

SEC.2II NEUTRON-PROTON SCATTERING

For a neutron-proton interaction describable by a central potential, the wave function in the center of mass system takes the form, for large separation, $r = r_N - r_P$, between the particles,

$$\emptyset \approx e^{i\underline{p}\cdot\underline{r}} + f(\theta)e^{ipr}/r. \tag{2.06}$$

The differential scattering cross section is related to the scattering amplitude, $f(\theta)$, by

$$d\sigma = |f(\theta)|^2 d\Omega. \tag{2.07}$$

In the low energy regime, where only s wave scattering is important, f is independent of angle and can be written

$$f(\theta) = (1/2ip)(e^{2i\delta} - 1), \tag{2.08}$$

where δ is the s wave phase shift. Eq.(2.08) can also be written

$$f = (1/p)e^{i\delta}\sin\delta \tag{2.09}$$
$$= (1/p)\sin\delta/e^{-i\delta} = (1/p)\sin\delta/(\cos\delta - i\sin\delta)$$
$$= 1/(p\cot\delta - ip). \tag{2.10}$$

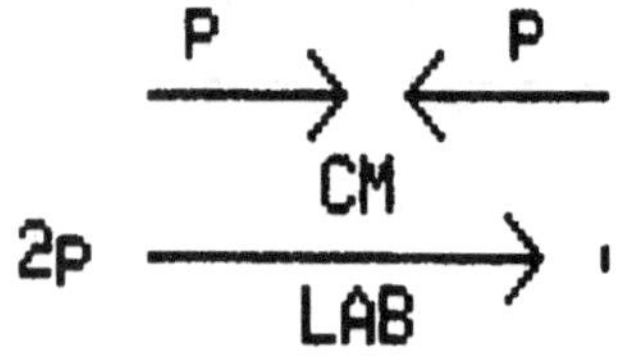

Since the scattering is spherically symmetric, the total cross section for elastic scattering is simply

$$\sigma = 4\pi|f|^2 = 4\pi/(p^2\cot^2\delta + p^2). \tag{2.11}$$

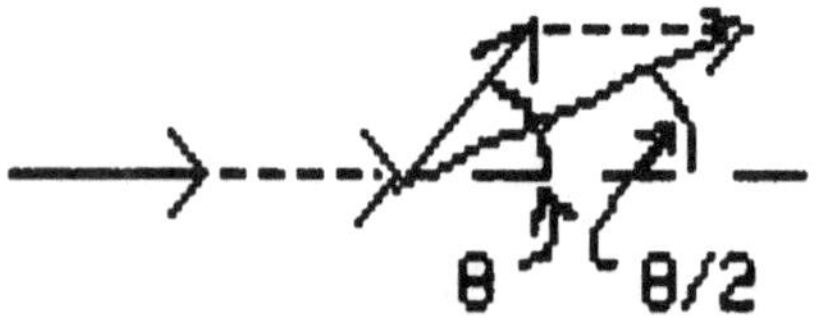

As illustrated in Fig.2.3, the scattering angle in the laboratory system (proton initially at rest) is half that in the center of mass system, $\theta_{lab} = \theta/2$, and

Fig.2.3

$$d\Omega = \sin\theta\, d\theta\, d\emptyset = 2\sin2\theta_{lab}\, d\theta_{lab}\, d\emptyset = 4\cos\theta_{lab}\sin\theta_{lab}\, d\theta_{lab}\, d\emptyset$$
$$= 4\cos\theta_{lab}\, d\Omega_{lab}.$$

Thus in the laboratory system

$$d\sigma = 4|f|^2\cos\theta_{lab}d\Omega_{lab}. \qquad (2.12)$$

Note that $0 \le \theta_{lab} \le \pi/2$.

The phase shift produced by a given potential, $V(r)$, can be found by solving the radial wave equation

$$u'' + (p^2 - MV(r))u = 0, \qquad (2.13)$$

where $p^2 = ME$, with E the energy in the center of mass system. For large r the potential $V(r)$ vanishes and the general solution of (2.13) is

$$u \approx c\sin(pr+\delta). \qquad (2.14)$$

The only requirement is that $V(r)$ fall off faster than $1/r$, as can easily be checked by using the WKB approximation. To evaluate δ it is necessary to integrate the wave equation, starting at the origin with $u=0$.

For the case of only s wave, that is, spherical scattering, it is easy to establish the identity of the δ appearing in (2.14) and that in (2.08). If there were no scattering, $V=0$, the wave function, call it $\phi^{(0)}$, would be a plane wave,

$$\phi^{(0)} = \exp^{i\underline{p}\cdot\underline{r}}. \qquad (2.15)$$

By assumption, only the spherically symmetric part of $\phi^{(0)}$ is affected by the scattering. The spherically symmetric part is obtained by averaging $\phi^{(0)}$ over the angles; by definition this gives the radial function $u^{(0)}/r$. Hence (with $\mu=\cos\theta$)

$$u^{(0)} = r\langle\phi\rangle = r\int e^{i\underline{p}\cdot\underline{r}}d\Omega/4\pi = (1/2)r\int_{-1}^{1}e^{ipr\mu}d\mu$$
$$= (1/2ip)(e^{ipr}-e^{-ipr}) = \sin(pr)/p. \qquad (2.16)$$

Evidently $u^{(0)}$ is the solution of the radial wave equation for $V=0$ which satisfies the boundary condition at the origin.

For a scattering problem, the incoming wave must be the same as that in (2.16); only the outgoing wave is

affected by the scattering. If (2.14) is written in terms
of incoming and outgoing waves,

$$u \approx (c/2i)(e^{i(pr+\delta)} - e^{-i(pr+\delta)}),$$

comparison with (2.16) shows that $c = e^{i\delta}/p$, and

$$u \approx (1/2ip)(e^{2i\delta}e^{ipr} - e^{-ipr}). \tag{2.17}$$

The difference between the outgoing waves in (2.17) and
(2.16) represents the scattered wave, so

$$\emptyset \approx e^{i\underline{p}\cdot\underline{r}} + (1/2ip)(e^{2i\delta} - 1)e^{ipr}/r,$$

and comparison with (2.06) yields (2.08).

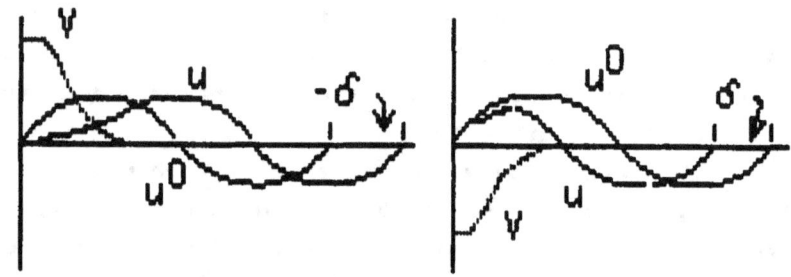

Fig.2.4

Fig.2.4 illustrates that an attractive potential, by
shortening the wavelength, produces a positive phase shift,
while a repulsive potential, which lengthens it or even
gives a reverse curvature, produces a negative phase shift.

Eqs. (2.08) and (2.14) only define δ modulo π, however
meaning can be given to δ itself. Consider the scattering
of a very low energy particle, so its wavelength is very
large compared to the range of the potential, by an
attractive potential whose magnitude is gradually increased
from zero. For V=0 take $\delta=0$. As the potential increases the
wavelength in the well shortens; at a certain magnitude the
slope of the wave function becomes zero just outside the
well. This is the point at which, according to (2.03) a

bound state appears at zero energy. At this point the phase
shift is $\pi/2$ (cos solution outside the well, rather than
sin). As shown in Fig.2.5, the outside solution reverses

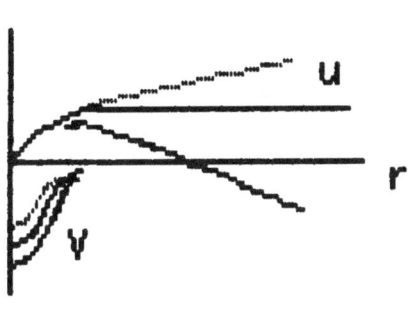

sign in going from a somewhat
smaller to a somewhat larger
potential. Thus the phase
shift increases by π when a
state becomes bound. If the
potential is further increased
till a second state becomes
bound, the phase shift will
again increase by π, etc. When
the magnitude of the potential
is not near a critical value
which produces a zero energy

Fig.2.5

bound state the magnitude of δ mod π is small, of order
pR, where R is the range. Thus for zero energy scattering
$\delta=n\pi$, where n is the number of bound states, reflecting the
fact that the wavefunction has n nodes. An equivalent
definition, and one applicable to repulsive potentials as
well, is to require δ to be a continuous function of E and
to approach zero as as E approaches infinity.

SEC.2III THE SCATTERING LENGTH

Fermi defined the scattering length, a, as the
negative of the zero energy scattering amplitude. Thus, for
zero energy,

$$f = -a, \quad \sigma = 4\pi a^2.$$

We see from (2.10) that when $p \to 0$, $p \cot \delta = -1/a$, so
$\cot \delta = -1/pa$ and $\delta \mod \pi = -pa$.

Fermi pointed out that the scattering length has a
simple geometrical meaning in terms of the wave function.
As $p \to 0$, (2.14), with $c = e^{i\delta}/p$, $\delta = n\pi - pa$, approaches
$u \approx (-1)^n (r-a)$. If we call the asymptotic form of $u(r)$, for
large r, $U(r)$, so

$$u(r) \approx U(r), \tag{2.18}$$

then $U(r)$ can be continued to small r. The point where
$U(r) = 0$ is the scattering length. Several cases are
illustrated in Fig.2.6: (a) shows the case for a repulsive

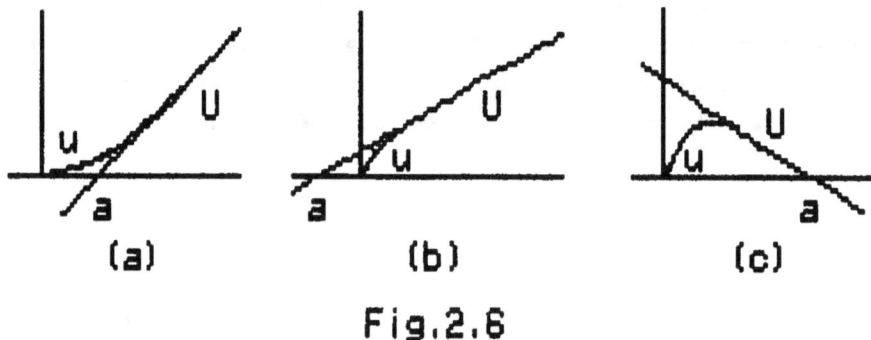

(a)   (b)   (c)

Fig.2.6

potential, (b) for an attractive potential with no bound
state and (c) for an attractive potential with one bound
state. Note that a is positive for a repulsive potential and
for an attractive potential with one bound state (as in the
case of the deuteron). For an attractive potential not deep

enough to give a bound state a is negative. The critical
case of a bound state of zero energy, where the slope of U
is zero, gives a=-∞.

SEC.2IV THE ZERO RANGE APPROXIMATION

For a potential of given shape the requirement that the scattering length have a certain value establishes a relation between the magnitude and range of the potential similar to that discussed in Sec.2I for a fixed bound state energy. The shorter the range the larger the potential that is required. If the range is very small compared to the scattering length, knowledge of the latter is sufficient to determine the scattering at finite energies.

We have used the term "range" rather loosely; let us suppose that it is defined so that at 'the the range, R, the wave function is already well represented by its asymptotic form, $u(R)=U(R)$. Let us further suppose that R is sufficiently small that $pR<<1$ for the energies under consideration. If this is so, then for $r<R$ the p^2 term in the wave equation, (2.13), is negligible and the wave function in the well is, to a good approximation, independent of energy. This becomes evident if one writes the wave equation in terms of a variable $y=r/R$ and considers the interval $0 \leq y \leq 1$. The wave function becomes

$$d^2u/dy^2+M(R^2V+p^2R^2)u=0.$$

To maintain a fixed scattering length, R^2V must remain finite as $R \rightarrow 0$, whereas the energy dependent term vanishes.

For $p=0$, $U(r)$ is proportional to $(r-a)$, and

$$u'(R)/u(R)=U'(R)/U(R)=-1/(a-R)\approx-1/a \qquad (2.19)$$

(the proportionality factor canceling in the ratio), since we have supposed $R<<a$. For finite energy $U(r)$ is given by (2.14) and

$$u'(R)/u(R)=U'(R)/U(R)=pcot(pR+\delta)\approx pcot\delta,$$

since $pR<<\delta$ (the critical case is near $p=0$, where δ $mod\pi$ also vanishes. However as $p \rightarrow 0$, δ $mod\pi \rightarrow -pa$, so the condition is again $R<<a$). Since the wave function inside the well is energy independent, so also is the value of

u'(R)/u(R), and we conclude that

$$p\cot\delta = -1/a. \tag{2.20}$$

We then have, from (2.10),

$$f = 1/[-(1/a)-ip] = -a/(1+ipa),$$

$$\sigma = 4\pi a^2/(1+p^2a^2). \tag{2.21}$$

The argument can be carried further if there is a bound state, as for the deuteron. The argument that the wave function for r<R is is independent of energy is as true for negative as for positive energies; the condition equivalent to pR<<1 is βR<<1, where

$$\beta = \surd(M\epsilon_D) = 1/4.318 \ f \tag{2.22}$$

for $\epsilon_D = 2.225$ Mev. The asymptotic deuteron wave function is, according to (2.03), $U = Ce^{-\beta r}$, so

$$u'(R)/u(R) = U'(R)/U(R) = -\beta.$$

Comparing with (2.19) determines the scattering length,

$$a = 1/\beta, \tag{2.23}$$

and gives a cross section

$$\sigma = (4\pi/\beta^2)/(1+p^2/\beta^2) = (4\pi/\beta^2)/(1+E/\epsilon_D). \tag{2.24}$$

SEC.2V THE SINGLET SCATTERING

At zero energy (2.24) would give a cross section $\sigma=2.34$ b (1 barn $=10^{-24}$ cm^2). The first cross section measurements[1], in 1935, gave an answer an order of magnitude larger, $\sigma\approx30$ b. The explanation of this large discrepancy was given by Wigner. Eq.(2.24) is based on the deuteron binding energy and thus gives the predicted scattering in the triplet state (S=1). The singlet state (S=0) of the N-P system must feel a different potential, since no bound singlet state exists, and must have a different scattering length. For an unpolarized neutron beam the relative probabilities of the neutron and the proton finding themselves in a triplet or a singlet state are proportional to the statistical weights of the states, so the average cross section will be

$$\sigma=(3/4)\sigma_t+(1/4)\sigma_s. \qquad (2.25)$$

The best present value of the epithermal cross section, based on measurements at lab energies between 6 and a few hundred electron volts[2] is $\sigma=20.442\pm0.023$ b. Using this value of σ and the calculated σ_t we find $\sigma_s=74.75$ b; equating this to $4\pi a_s^2$ gives $a_s=-24.34$ f. (The cross section determines only a_s^2; the negative sign is taken because there is no bound singlet state.) This value is surprisingly large, improbable but not impossible; it means that the depth of the singlet well must be near the critical value which would produce a bound state.

The value of a_s given above is not quite right, because of the inaccuracy of the zero-range approximation

[1] T.Bjerge and C.H. Westcott, Proc. Roy. Soc. 150A, 709. (1935); J.R. Dunning,G.B. Pegram, G.A. Fink and D.P. Mitchell, Phys. Rev.47, 910 (1935).
[2] L.Koester, Zeit. Phys., 198, 187 (1967)

which was used to estimate r_t. The best value of a_s is

$$a_s = -23.715 \pm 0.013 \text{ f.} \qquad (2.26)$$

The values of f^2 needed for a Yukawa potential to give this scattering length are shown in Table 2.3 for a choice of ranges. Remember that $f^2 = 1.68$ gives a zero energy bound state. Table 2.4 is for the more realistic case of a Yukawa potential outside a repulsive core of radius 0.5 f.

Table 2.3			
$1/\mu$	f^2	$g^2/4\pi$	r_s
0	1.68	∞	0
0.70	1.61	0.485	1.56
1.05	1.58	0.317	2.41
1.40	1.55	0.233	3.29

Table 2.4			
$1/\mu$	f^2	$g^2/4\pi$	r_s
0	∞	∞	1.02
0.35	17.72	10.64	2.07
0.70	6.16	1.85	3.03
1.05	4.10	0.82	3.99
0.60	8.49	2.97	2.76

Equation (2.25) gives for the scattering cross section in zero range approximation

$$\sigma = 4\pi[(3/4) a_t^2/(1+p^2 a_t^2) + (1/4) a_s^2/(1+p^2 a_s^2)]$$
$$= (3/4) \times 2.34/(1+E/\epsilon_D) + (1/4) \times 74.75/(1+E/\epsilon_s).$$

Here we have defined an energy $\epsilon_s = 1/Ma_s^2 = 70$ Kev for $a_s = -24.34$ f., the zero-range value. If a_s were positive, rather than negative, ϵ_s would be the binding energy of the singlet state. The light curve in Fig. 2.7 shows the values predicted by this expression. The measured values of the cross section, for lab energies between 35 Kev and 6 Mev, have a 5% accuracy, which it was possible to achieve in 1946 by using mono-energetic protons and deuterons from a Van der Graff accelerator to produce mono-energetic

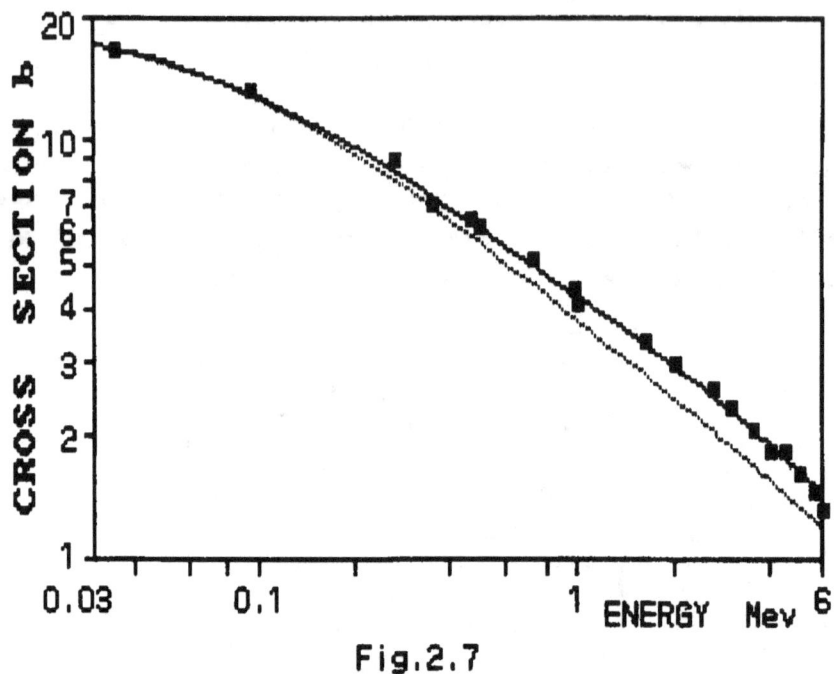

Fig.2.7

neutrons from light element reactions.[3] At energies above
1 Mev, where the cross section is mostly due to triplet
scattering, the calculated values are about 20% too low.
This is because the zero-range approximation, which gives
a_t=4.317 f, is inadequate. The actual value is

$$a_t=5.4255\pm0.004 \text{ f}. \tag{2.27}$$

The values given by (2.26) and (2.27) satisfy the zero-
energy scattering relation; at zero energy

$$\sigma=\pi(3a_t^2+a_s^2). \tag{2.28}$$

[3] C.L. Bailey, W.E.Bennett, T. Bergstralh, R.G. Nuckolls,
H.T. Richards and J.H. Williams, Phys. Rev.70, 583 (1946);
D.H. Frisch, Phys. Rev.70, 589 (1946).

SEC.2VI THE EFFECTIVE RANGE

Before 1948 much effort was expended in trying to deduce the shape of the N-P potential from data on scattering such as that illustrated in Fig.2.7. Much of this effort was misguided because, as was pointed out by Landau and Smorodinsky in 1944 and later by Schwinger in 1947[4], all that can be deduced from the scattering in the energy range below 10 Mev is, in addition to the scattering lengths, two other parameters related to the ranges of the triplet and singlet forces. For any shape potential, knowledge of the scattering length and "effective range" determines the magnitude of the potential and its range. But any short range potential will do equally well. We present the argument in the form given by Chew and Goldberger in 1949[5].

It will be convenient, for this argument, to choose the multiplicative constant in (2.14) to be $c=1/\sin\delta$, so

$$U(0)=1. \tag{2.29}$$

We shall distinguish the wave functions for different energies by attaching the subscript p to u and U. Thus

$$U_p=\sin(pr+\delta)/\sin\delta, \tag{2.30}$$
$$U_0=(1-r/a), \tag{2.31}$$

remembering that, for zero energy, U vanishes at $r=a$. The idea of the calculation is that the kinetic energy term in the wave equation can be regarded as small within the range of the potential, so one seeks an expansion of $p\cot\delta$ in powers of the energy, i.e. of p^2. Let us write the wave equation with the "small" term on the right,

$$u_p''-MVu_p=-p^2u_p, \quad u_0''-MVu_0=0.$$

[4] L. Landau and J. Smorodinsky, J. Phys.(U.S.S.R.)8,154 (1944); J.S. Schwinger, Phys. Rev.72,742 (1947).
[5] G.F Chew and M.L. Goldberger, Phys. Rev., 75, 1637 (1949).

Multiply the first equation by u_0, the second by u_p, and subtract. The result can be written

$$(d/dr)W(u_0,u_p)=-p^2 u_0 u_p,$$

where W is the Wronskian,

$$W(u_0,u_p)=u_0 u_p' - u_0' u_p.$$

Integrating from zero to an arbitrary r_1, we have

$$W(u_0,u_p)\Big|_0^{r_1} = -p^2 \int_0^{r_1} u_0 u_p dr.$$

Since U_p, U_0 satisfy the same wave equation with V=0, we also have

$$W(U_0,U_p)\Big|_0^{r_1} = -p^2 \int_0^{r_1} U_0 U_p dr,$$

and

$$[W(U_0,U_p)-W(u_0,u_p)]\Big|_0^{r_1} = -p^2 \int_0^{r_1}(U_0 U_p - u_0 u_p)dr.$$

Since $u \to U$ at large r, we can allow $r_1 \to \infty$ in this expression: on the left the two Wronskians cancel at the upper limit, and the integral on the right converges. Moreover, at the lower limit, $W(u_0,u_p)_0=0$, since $u_0(0)$, $u_p(0)$ are both zero. We are left with

$$W(U_0,U_p)_0 = p^2 \int_0^\infty (U_0 U_p - u_0 u_p)dr. \qquad (2.32)$$

Using (2.30) and (2.31) we find $W(U_0,U_p)_0 = p\cot\delta + 1/a$, and

$$p\cot\delta = -1/a + p^2 \int_0^\infty (U_0 U_p - u_0 u_p)dr. \qquad (2.33)$$

The expression (2.33) is exact, no approximations were made in deriving it. However, if we suppose that $p\cot\delta$ can be expanded in a power series in p^2, the coefficient of the p^2 term can be found by neglecting the p dependence of the integral, that is, by replacing U_p and u_p by U_0 and u_0. We then have the approximate relationship

$$p\cot\delta = -1/a + p^2 \int_0^\infty (U_0^2 - u_0^2)dr. \qquad (2.34)$$

To put the argument more physically, we note that the contributions to the integral in (2.33) come from within the range, as defined in Sec.2IV, and that, as argued in Sec.2IV, the wave functions are nearly energy independent for $0<r<R$, which justifies the approximation.

The integral which appears in in (2.34) has the dimensions of a length. We accordingly define an "effective range", r_e, by

$$r_e/2 = \int_0^\infty (U_0^2 - u_0^2) dr, \qquad (2.35)$$

and write (2.34) in the conventional form

$$p\cot\delta = -1/a + (1/2)r_e p^2. \qquad (2.36)$$

As as example, consider a rather extreme case, hard sphere scattering ($V=\infty$, $r<a$; $V=0$, $r>a$). Here $U_0 = 1 - r/a$; $u_0 = 0$, $r<a$; $u_0 = U$, $r>a$, as shown in Fig.2.8, and

$$r_e = 2\int_0^a (1-r/a)^2 dr = (2/3)a.$$

On the other hand, $u_p = c \cdot \sin p(r-a)$, so $\delta = -pa$. The exact relationship is

$$p\cot\delta = p\cot(-pa) = -1/a + (1/3)ap^2 + (1/45)a^3p^4 + \cdots.$$

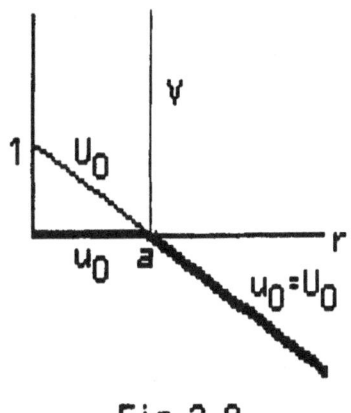

Fig.2.8

The effective range approximation will be seen to give the p^2 term correctly. The smallness of the coefficient of the a^3p^4 term is noteworthy. Even for $ap=1$, neglect of this term makes only a 2% error in the cross section. Calculations with various short range potentials give this as a characteristic feature. Before the experimental techniques were refined enough to permit 1% scattering experiments, experiments with 100 Mev neutrons became possible at the Berkeley 184" cyclotron in 1949, and observations of the angular distribution of the scattering, as well as its magnitude, began to give more detailed information on the shape of nuclear forces.

Just as in the case of the zero-range approximation discussed in Sec.2IV, the expression (2.32) also holds for negative energies, that is, for the case of a bound state near zero energy. All that is necessary, for the deuteron case, is to replace p^2 by $-\beta^2$, and, of course, U_p and u_p by U_D and u_D. Since, with the normalization (2.29), $U_D = e^{-\beta r}$, we obtain $W(U_0, U_D) = -\beta + 1/a_t$, and (2.32) gives

$$1/a_t = \beta - \beta^2 \int_0^\infty (U_0 U_D - u_0 u_D) \, dr. \qquad (2.37)$$

With the same approximation as before, U_D and u_D can be replaced by U_0 and u_0 in the integral, and we find

$$1/a_t = \beta - (1/2)\beta^2 r_t, \qquad (2.38)$$

where r_t is the triplet effective range. The first term on the right corresponds to the zero-range approximation, (2.23), the second gives a correction due to the finite range. If we use the values of a_t and β given by (2.27) and (2.22) we find

$$r_t = 1.763 \pm 0.005 \ f. \qquad (2.39)$$

These relations are illustrated in Fig.2.9. The line

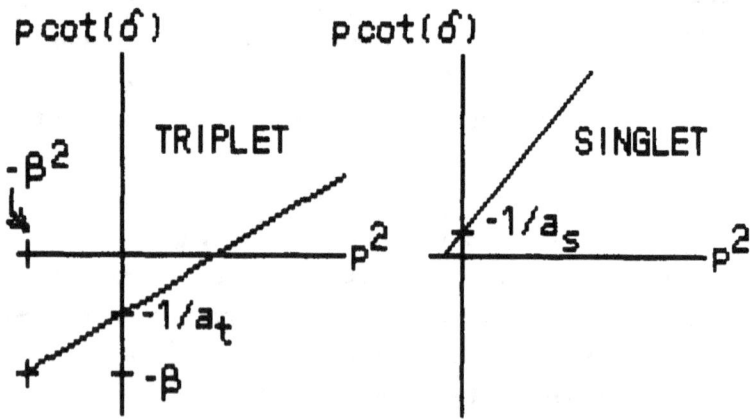

Fig.2.9

for the triplet state is determined by the two points at
$p^2=0$ and $p^2=-\beta^2$, the bound state energy. Note that the
phase shift is $\delta=\pi$ at zero energy, and decreases with
increasing energy. It equals $\pi/2$ when $p\cot\delta=0$, i.e. when
$p^2=2/r_t a_t$. With the quoted values, this gives $\delta=\pi/2$ at
a center of mass energy E=8.67 Mev. The singlet phase shift
is zero at zero energy, rises to a maximum at $p^2=-2/r_s a_s$,
then decreases.

The determination of r_s (the slope of the singlet
curve in Fig.2.8) requires highly accurate measurements of
the scattering cross section. In 1963 Engelke et al.[*] made
such measurements at two energies, and found σ=6.202±0.18% b
at 0.4926 Mev and σ=2.206±0.31% b at 3.205 Mev. A critical
evaluation by H.P. Noyes and H. Fiedeldey[7], using these
and other measurements
at the 1% level, gave r_s=2.76±0.07 f.

Collecting the values:

$$a_t=5.425\pm0.004 \text{ f},$$
$$r_t=1.763\pm0.005 \text{ f},$$
$$a_s=-23.715\pm0.013 \text{ f}$$
$$r_s=2.76\pm0.07 \text{ f}. \tag{2.40}$$

The scattering cross sections calculated using the
effective range formulas (2.11), (2.25), (2.36), and the
values (2.40) are shown by the heavy curve in Fig.2.7.

Table 2.1 for Yukawa potentials and Table 2.2 for
Yukawa potentials with a repulsive core show the calculated
values of a_t and r_t for various ranges. The difference
between r_t calculated from (2.38) and from (2.35) gives a
measure of the accuracy of the effective range
approximation. The range which agrees with (2.40) is shown
in the last line of Table 2.2. Data for the singlet state

[*] C.E. Engelke, R.E. Benenson, E. Melkonian and J.M.
Lebowitz, Phys. Rev., 129, 324 (1963).
[7] Quoted by E.M. Henley, Isospin in Nuclear Physics, D.H.
Wilkinson, editor, North Holland Publishing Co. (1969).

is given in Tables 2.3 and 2.4. When the repulsive core is included the ranges correspond more closely to two pion exchange than to one.

CHAPTER 3

LOW ENERGY
NEUTRON–PROTON SCATTERING

SEC.31 EFFECT OF CHEMICAL BONDING

The zero-energy scattering amplitudes, whose values
were quoted without explanation in the previous chapter,
are, in fact, determined by a variety of scattering
experiments at very low energy, usually using "thermal
neutrons", which have a roughly Maxwellian velocity
dependence after being slowed by collisions in hydrogenous
materials. For room temperature, kT=1/40 ev.

Fermi and Amaldi[1] measured the cross section for
scattering of thermal neutrons by protons bound in
hydrogenous materials and found a cross section of 86 b.
Later experiments[2] using monochromatic slow neutrons gave
scattering cross sections rising to 80 b for very slow
neutrons, four times the value given in the previous
chapter. That value holds for epithermal neutrons, that is
for neutrons of energy over 1 ev. That the binding of the

---

[1] E. Amaldi and E. Fermi, Ricerca Scientifica 1, 223 (1936)
[2] W.B. Jones, Phys. Rev., 74, 364 (1948).

proton in a molecule could affect the scattering at first
sight seems surprising, because of the short range and
large magnitude of the nuclear force, but it must be
remembered that if the neutron energy is less than the
energy required to excite the proton to its first
vibrational state in the molecule then energy conservation
requires the proton to remain in the ground vibrational
state after the collision. As a result, the molecule
recoils from the collision as a whole and the kinematics
are quite different than in a free collision. Typical
molecular vibrational energies are a few tenths of an
electron volt. Fermi[3] showed that, in the limiting case
of low energy and heavy molecules,

$$\sigma_{bound} = 4\sigma_{free}.$$

Let us consider the collision in the laboratory
system, where the incoming neutron has momentum p and the
molecule is at rest. Let $v_n(r_P)$ be the proton wave function
in the nth vibrational state, $v_0(r_P)$ the ground state wave
function. The cross section for scattering leaving the
molecule in the nth excited state (provided this is
energetically possible) and giving a scattered neutron of
momentum p_n in $d\Omega$ is

$$d\sigma_n = (2\pi/v)|V_{n,p_n}|^2 p_n^2 d\Omega / [(2\pi)^3 (dE/dp_n)], \qquad (3.01)$$

where

$$V_{n,p_n} = \int e^{-ip_n \cdot r_N} v_n(r_P)^* V(|r_N - r_P|) \Xi(r_N, r_P) dr_N dr_P. \qquad (3.02)$$

In this relationship, which is exact, $V(|r_P - r_N|)$ is the N-P
interaction and $\Xi(r_P, r_N)$ is the exact wave function.
Usually, in this type of problem, Ξ is not known exactly
and some approximation is chosen, e.g. the Born
approximation takes $\Xi = \exp(ip \cdot r_N) v_0(r_P)$. The energy relation

---

[3] E. Fermi, Ricerca Sci., 7, 364 (1936).

is

$$p^2/2M = E = p_n^2/2M + (\underline{p}-\underline{p}_n)^2/2M_A + n\omega$$
$$= p_n^2/2M + (p^2-2pp_n\cos\theta+p_n^2)/2M_A + n\omega, \qquad (3.03)$$

with M_A the mass of the molecule and ω the vibrational frequency. The fact that $\cos\theta$ appears linearly in (3.03), whereas p_n^2 appears quadratically, suggests that it would be a little easier, when evaluating the density of states per unit energy, to eliminate $d\cos\theta$ rather than dp_n. In getting (3.01) one wrote

$$d\underline{p}_n/dE = p_n^2 \ (dp_n/dE)d\emptyset d\cos\theta.$$

One could equally take

$$d\underline{p}_n/dE = p_n^2 dp_n d\emptyset |d\cos\theta/dE|,$$

and obtain

$$d\sigma_n = (2\pi/v)|V_{n,\underline{p}_n}|^2 p_n^2 dp_n d\emptyset/[(2\pi)^3|dE/d\cos\theta|], \qquad (3.04)$$

the cross section for scattering into $dp_n d\emptyset$. From (3.03), $|dE/d\cos\theta| = pp_n/M_A$, so, since $v=p/M$,

$$d\sigma_n = (MM_A/4\pi^2)|V_{n,\underline{p}_n}|^2 p_n dp_n d\emptyset/p^2. \qquad (3.05)$$

Relations analogous to (3.01) and (3.02) can be written for free N-P scattering, evaluated in the center of mass system. Here we have, if the initial momentum is $\underline{p}$ and the final one $\underline{p}'$,

$$d\sigma_{free} = (2\pi/v)|V_{\underline{p}'}|^2 p'^2 d\Omega/[(2\pi)^3(dE/dp')],$$

with $V_{\underline{p}'} = \int e^{-i\underline{p}'\cdot\underline{r}} V(r)\underline{g}(\underline{r})d\underline{r}$. The energy relation, in terms of the reduced mass, $\mu=M/2$, is $p^2/2\mu = E = p'^2/2\mu$, so $p'=p$ and $dE/dp'=v=p/\mu$. Thus

$$d\sigma_{free} = (\mu^2/4\pi^2)|V_{\underline{p}'}|^2 d\Omega. \qquad (3.06)$$

The scattering amplitude is

$$f_{free} = -(\mu/2\pi)V_{\underline{p}'}, \qquad (3.07)$$

the negative sign being needed because a positive V gives a negative f. In this case $\underline{g}(r) = u(r)/r$, with $u(r)$ normalized as in (2.14) and (2.17), and we have an exact relationship

$$f_{free} = -(\mu/2\pi)\int e^{-i\underline{p}'\cdot\underline{r}}V(r)u(r)d\underline{r}/r.$$ (3.08)

We are now prepared to make a reasonable guess at the $\S(\underline{r}_N,\underline{r}_P)$ appearing in (3.02). Note that is only necessary to know $\S$ where $V(r)$ differs from zero. We take

$$\S(\underline{r}_N,\underline{r}_P) = e^{i\underline{p}\cdot(\underline{r}_N+\underline{r}_P)/2}[u(|\underline{r}_N-\underline{r}_P|)/|\underline{r}_N-\underline{r}_P|]v_0(\underline{r}_P).$$ (3.09)

Since the calculation is being done in the laboratory system, rather than the N-P center of mass system, we include the plane wave factor representing the motion of the N-P center of mass with momentum $\underline{p}$. Equation (3.09) expresses the feeling that the weak long range molecular forces cannot affect the scattering process: it says that when neutron and proton are within the range of the nuclear force the wave function is the same as in free scattering. If (3.09) is substituted in (3.02), and a shift made in the origin of the $\underline{r}_N$ integration, to $\underline{r}_N=\underline{r}+\underline{r}_P$, we obtain a product of two factors

$$V_{n,\underline{p}_n} = \int v_n(\underline{r}_P)^* e^{i\underline{p}_P\cdot\underline{r}_P}v_0(\underline{r}_P)d\underline{r}_P \times \int \exp(-i\underline{p}_n'\cdot\underline{r})V r)u(r)d\underline{r}/r,$$ (3.10)

where $\underline{p}_P=\underline{p}-\underline{p}_n$ is the momentum the recoil proton would have if the momentum of the N-P system were conserved in the collision, and $\underline{p}_n'=\underline{p}_n-(1/2)\underline{p}$ is the momentum of the scattered neutron in the N-P center of mass system. The first factor is called the form factor, and will be denoted $F_n(\underline{p}_P)$.

Let us first consider elastic scattering (n=0). The form factor is

$$F_0(\underline{p}_P) = \int e^{i\underline{p}_P\cdot\underline{r}_P}|v_0(\underline{r}_P)|^2 d\underline{r}_P.$$

For forward neutron scattering $\underline{p}_P=0$, and $F_0=1$. The proton density in the molecule, $|v_0(\underline{r}_P)|^2$, has a spread given by the zero-point amplitude of the oscillator, x_0. For a harmonic oscillator $E_0=w/2=(1/2)Mw^2x_0^2$ so $w=1/Mx_0^2$. If $p_P^2/2M=w/2$ we have $p_P x_0=1$. The form factor F_0 falls off

rapidly when $p_P x_0 > 1$, thus when the proton recoil energy
becomes comparable to w.

The second factor in (3.10) is the same integral that
occurs in (3.08). Because of the different kinematics of
free and bound collisions the range of the variable $\underline{p}_0'$ is
different than $\underline{p}'$. However, as long as $p_0' R \ll 1$ (with R the
range of $V(r)$) the exponential factor equals one in both
integrals and is irrelevant. Equation (3.05) becomes

$$d\sigma_0 = (M M_A/\mu^2) |F_0(\underline{p}_P)|^2 |f_{free}|^2 p_0 dp_0 d\phi/p^2. \qquad (3.11)$$

Let us take the case where the incident neutron energy
is less than w, so only elastic scattering is possible, and
further, sufficiently small compared to w that $|F_0(\underline{p}_P)|^2 = 1$
for all angles of scattering. The total cross section is
then

$$\sigma_0 = 2\pi (M M_A/\mu^2) |f_{free}|^2 \times (1/2) (p_{0,max}^2 - p_{0,min}^2)/p^2.$$

The maximum value of p_0 is obtained for forward scattering,
where $p_{0,max} = p$. It can easily be checked from the energy
relation (Equation (3.03) with n=0) that the minimum, for
backward scattering, is $p_{0,min} = [(M_A - M)/(M_A + M)]p$. A little
algebra gives

$$\sigma_0 = 4\pi (\mu_A/\mu)^2 |f_{free}|^2, \qquad (3.12)$$

where μ_A is the N-molecule reduced mass, $\mu_A = M M_A/(M + M_A)$. For
a molecule of atomic weight A,

$$(\mu_A/\mu)^2 = 4A^2/(A+1)^2. \qquad (3.13)$$

Thus for large A the scattering cross section is four times
the free cross section, as is in fact observed for very
slow neutrons.

In order to understand the mysterious appearance of
the factor $(\mu_A/\mu)^2$ in (3.12), let us go back to (3.01) and
look at the elastic cross section per unit solid angle in
the forward direction. In the forward direction we find
from (3.03) $dE/dp_0 = p/M$, and (3.01) for $d\sigma(\theta)$ becomes

$$d\sigma(0) = (M^2/4\pi^2) |V_{0,\underline{p}}|^2 d\Omega. \qquad (3.14)$$

Whatever the neutron energy, $p_P=0$, $F_0(0)=1$, and, since $p_0=p$, the magnitude of p'_0 in (3.10) is the same as p' in (3.08). Thus

$$d\sigma(0)=(M/\mu)^2|f_{free}|^2 d\Omega=4|f_{free}|^2 d\Omega. \qquad (3.15)$$

We see from (2.12) that in the laboratory system the forward scattering amplitude for free scattering is

$$f_{free}(0)=2f_{free}. \qquad (3.16)$$

Thus (3.15) tells us that if both scattering amplitudes are given in the lab system

$$f_{bound}(0)=f_{free}(0). \qquad (3.17)$$

The forward scattering amplitude is unaffected by the binding. (As we shall see later in this chapter, the forward elastic scattering amplitude in the laboratory system has a direct physical meaning in terms of the mean potential the neutron feels in passing through a scattering medium.) The reason the cross section is different is because the angular distributions in the laboratory system are different. For free scattering the the angular distribution is $\cos\theta$ in the forward hemisphere, while for large A it is spherically symmetric, a difference which just accounts for the factor four.

There is another way to interpret (3.12). In a scattering problem, consider the probability of elastic scattering within a cone of small angle θ around the forward direction. This is $|f(0)|^2 \cdot \pi\theta^2$. In terms of the transverse momentum of the scattered particle, $p_{tr}=p\theta$, the probability of scattering with transverse momentum less than p_{tr} is $(|f(0)|^2/p^2) \cdot \pi p_{tr}^2$. Since p_{tr} is unchanged by a Galilean transformation parallel to the initial momentum p, it follows that $f(0)/p$ is invariant under such a transformation. A particular case is

$$f(0)_{lab}=(p_{lab}/p_{CM})f_{CM}=(M/\mu')f_{CM}, \qquad (3.18)$$

where μ' is the reduced mass ($p_{lab}=Mv$, $p_{CM}=\mu'v$). Equation

(3.16) is an example of this rule. Equation (3.12) is
simply the result of applying (3.18) to both amplitudes in
(3.17).

As the neutron energy is
raised the elastic cross
section falls off as $|F_0|^2$
decreases. At the n=1
threshold the inelastic
scattering comes in, etc., as
illustrated in Fig. 3.1. For
E>>ω the cross section
approaches the free value, as
we shall now show. To avoid a
little kinematics, let us take
$M_A=\infty$, so the energy relation,

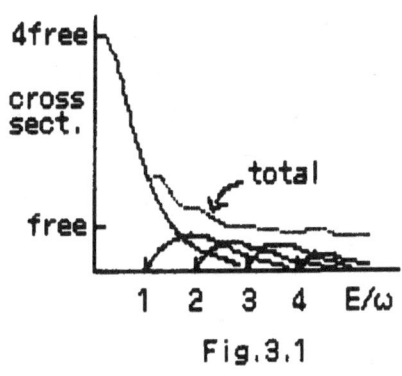

Fig.3.1

(3.03), becomes

$$E=p_n^2/2M + n\omega. \qquad (3.19)$$

The form factor

$$F_n(\mathbf{p_p})=\int v_n(\mathbf{r_p})^* e^{i\mathbf{p_p}\cdot\mathbf{r_p}} v_0(\mathbf{r_p})d\mathbf{r_p} \qquad (3.20)$$

has a simple interpretation. In the collision the proton is
suddenly (in a time short compared to 1/ω) given a momentum
$\mathbf{p}$. Before the collision its wave function was $v_0(\mathbf{r_p})$,
just after it is $\exp(i\mathbf{p_p}\cdot\mathbf{r_p})v_0(\mathbf{r_p})$. F_n is just the
probability amplitude for finding the proton with this wave
function in the state n. This is a standard problem in
quantum mechanics courses: A harmonic oscillator in its
ground state $v_0(x)$ is suddenly given a momentum p_p, so its
wave function becomes $\underline{g}(x,0)=\exp(ip_px)v_0(x)$. Find the
subsequent wave function, and the probability of being in
state n. The answer is that $|\underline{g}(x,t)|^2$ is a packet with the
same spatial shape as $|v_0(x)|^2$ which moves back and forth
in the potential well with simple harmonic motion, with the
amplitude appropriate to an oscillator of energy $p_p^2/2M$ (see
Fig. 3.2). The probability distribution

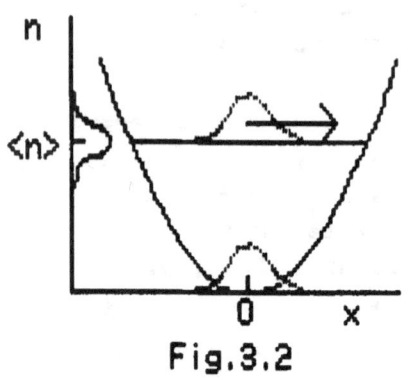

$$P_n=|F_n|^2=|\int v_n(x)^* e^{ip_P x} v_0(x)dx|^2$$

is a Poisson distribution in n, $P_n=e^{-\langle n\rangle}\langle n\rangle^n/n!$, with $\langle n\rangle$ given by

$$\langle n\rangle \omega = p_P^2/2M. \qquad (3.21)$$

We can easily check (3.21). The expectation value of the energy, for an oscillator with wavefunction $\exp(ip_P x)v_0(x)$, is

Fig.3.2

$$\langle E\rangle=\int v_0(x)^* e^{-ip_P x} He^{ip_P x} v_0(x)dx,$$

with $H=p^2/2M +(1/2)M\omega^2 x^2$, $p=(1/i)d/dx$. Since $p\exp(ip_P x)=\exp(ip_P x)(p+p_P)$, this gives

$$\langle E\rangle=\int v_0(x)^*[(p+p_P)^2/2M +(1/2)M\omega^2 x^2]v_0(x)dx=p_P^2/2M +\omega/2.$$

Setting $\langle E\rangle=(\langle n\rangle+1/2)\omega$ gives (3.21).

For large $\langle n\rangle$ the Poisson distribution has a sharp peak at $\langle n\rangle$, with a mean-square deviation $(\Delta n)^2=\langle n\rangle$. For $n=\langle n\rangle$ the energy relation (3.19) is, in virtue of (3.21),

$$E=p_{\langle n\rangle}^2/2M + p_P^2/2M. \qquad (3.22)$$

Remembering that, by definition, $p_P=p-p_{\langle n\rangle}$, we see that (3.22) is exactly the energy relation for free scattering. (The mean-square deviation of the scattered neutron's energy from the free value is, as we see from (3.19), $(\Delta E_n)^2=\omega^2(\Delta n)^2=\omega^2\langle n\rangle=\omega E_P$, with $E_n=p_n^2/2M$, $E_P=p_P^2/2M$.) From (3.01), (3.19) and (3.10)

$$d\sigma_n=(M^2/4\pi^2)|F_n|^2|\int \exp(-ip_n'\cdot r)Vr)u(r)dr/r|^2 p_n d\Omega/p.$$

The factors other than $|F_n|^2$ are slowly varying in n, and in them n may be replaced by $\langle n\rangle$. The integral is then exactly that appearing in (3.08), since for $\langle n\rangle$ the kinematics are identical with that of a free collision, and

$$d\sigma_n=4|F_n|^2|f_{free}|^2 p_{\langle n\rangle}d\Omega/p.$$

Summing over n

$$d\sigma=\Sigma_n d\sigma_n=4|f_{free}|^2 p_{\langle n\rangle}d\Omega/p,$$

the final form resulting from the sum rule $\Sigma_n |F_n|^2 = 1$.

Written as a matrix element $F_n = (e^{i\underline{p}_p \cdot \underline{r}})_{n0}$, and

$$\Sigma_n |F_n|^2 = \Sigma_n (e^{-i\underline{p}_p \cdot \underline{r}})_{0n} (e^{i\underline{p}_p \cdot \underline{r}})_{n0}$$
$$= (e^{-i\underline{p}_p \cdot \underline{r}} e^{i\underline{p}_p \cdot \underline{r}})_{00} = 1.$$

Finally, (3.22) gives, as for free scattering, $p_{\langle n \rangle} = p\cos\theta$, and

$$d\sigma = 4|f_{free}|^2 \cos\theta d\Omega,$$

the free scattering cross section in the laboratory system.

We have been a little cavalier in talking of the vibrational motion of the proton as if it were a one dimensional oscillator, whereas F_n is a volume integral. Actually, the proton coordinates should be expressed in terms of the normal modes of the molecule, and F_n should be regarded as a product of the form factors for each mode. If any mode has a null frequency it means we are dealing with a rotational, rather than a vibrational, mode of motion. The rotational splittings of molecules are of the order of 10^{-4} ev. For experiments which do not discriminate such small energy transfers, the argument about the quasi-elastic scattering which was just given for the vibrational modes can be applied as well to the rotational modes. The result is equivalent to forgetting the rotational modes, evaluating the cross section for a given orientation of the molecule, then averaging classically over orientations.

Also, we have considered only scattering from a single proton. In fact, all the nuclei in the molecule scatter the neutron. For the case described by (3.12), where the neutron's wavelength is sufficiently large that the vibrational form factors are one, we have

$$\sigma = 4\pi \langle |\Sigma_i \exp(i\underline{q}\cdot\underline{r}_i) \cdot (\mu_A/\mu_i) f_i|^2 \rangle, \tag{3.23}$$

the index i running over all nuclei in the molecule, and f_i

being the scattering amplitude of the ith nucleus. Here $q=p-p'$, with p and p' being the initial and scattered neutron momenta, and r_i is the equilibrium position of the the ith nucleus in the molecule. The exponential factors simply give the phase differences of the waves scattered from the various nuclei. The average is to be taken over all orientations of the molecule. If the wavelength is still short compared to internuclear separations the cross terms in the sum would vanish on averaging, and

$$\sigma = 4\pi \Sigma_i (\mu_A/\mu_i)^2 f_i^2.$$

If the wavelength were much larger than the molecule

$$\sigma = 4\pi |\Sigma_i (\mu_A/\mu_i) f_i|^2. \tag{3.24}$$

SEC.3II SCATTERING OF SLOW NEUTRONS BY HYDROGEN MOLECULES

The H_2 molecule, because of the low mass of the nuclei
and the small internuclear separation, 0.746 A, has an
unusually small moment of inertia and an unusually high
rotational energy splitting. The excitation of the J=1
state is $1.5 \cdot 10^{-3}$ ev. For comparison, at 20° K, the boiling
point of liquid hydrogen, $kT=1.7 \cdot 10^{-3}$ ev, and a neutron of
this energy would have $1/p \approx 1.1$ A. This circumstance permits
some interesting information to be obtainable from N-H_2
scattering, as was first pointed out by Teller and
Schwinger in 1937.[4]

The H_2 molecule exists in two forms, called para and
ortho hydrogen. The spins of the two protons can couple to
produce a resultant spin F=1 (orthohydrogen) or F=0
(parahydrogen). The requirement that the wavefunction of
the molecule must be antisymmetric under exchange of the
two protons produces a correlation between F and the
rotational quantum number J. For the H_2 wavefunction, a
reflection in the origin (center of mass) gives a factor
$(-1)^J$. Furthermore, the wavefunction is even under
reflection of the electrons alone in the origin, holding
the nuclei fixed, which is also a symmetry operation for a
diatomic molecule composed of two similar atoms. The
product of of the two operations, which simply interchanges
the coordinates of the protons, thus multiplies their
orbital wavefunction by $(-1)^J$. Antisymmetry then requires
J even for F=0 (odd spin function) and J odd for F=1 (even
spin function). At room temperature the ratio of para to
ortho hydrogen is the ratio of the statistical weights,
1:3. If the hydrogen is cooled to 20° this ratio remains
unchanged, because the ortho-para conversion goes very

---

[4] J.S. Schwinger and E. Teller, Phys. Rev., 52, 286
(1937).

slowly without a catalyst. After conversion by a catalyst the ratio is more nearly 1:1. By measuring the neutron scattering cross section before and after conversion one can thus determine both σ_{ortho} and σ_{para}.

The spin dependence of the low energy N-P scattering amplitude can be expressed by writing $f_{NP}=-(a_t P_t + a_s P_s)$, where a_t and a_s are the triplet and singlet scattering lengths and P_t and P_s are the projection operators for the triplet and singlet spin states. Since $\mathbf{s}_N \cdot \mathbf{s}_P = 1/4$ for the triplet state and $-3/4$ for the singlet state,

$$P_t = 3/4 + \mathbf{s}_N \cdot \mathbf{s}_P = \begin{cases} 1, & S=1 \\ 0, & S=0, \end{cases}$$

$$P_s = 1/4 - \mathbf{s}_N \cdot \mathbf{s}_P = \begin{cases} 0, & S=1 \\ 1, & S=0. \end{cases}$$

Thus

$$f_{NP} = -[(3/4)a_t + (1/4)a_s + (a_t - a_s)\mathbf{s}_N \cdot \mathbf{s}_P]. \qquad (3.26)$$

For a neutron whose wavelength is long compared to 0.746 A, the scattering amplitude of the molecule, in the N-H$_2$ center of mass frame, is (see (3.24) with $\mu_A = 2/3$, $\mu' = 1/2$)

$$f = -2(2/3)[(3/2)a_t + (1/2)a_s + (a_t - a_s)\mathbf{s}_N \cdot \mathbf{F}], \qquad (3.27)$$

with $\mathbf{F} = \mathbf{s}_{P1} + \mathbf{s}_{P2}$. For parahydrogen F=0, and

$$\sigma_{para} = 4\pi (4/3)^2 f_c^2,$$

where

$$f_c = -(1/2)(3a_t + a_s) \qquad (3.28)$$

is called the coherent scattering amplitude. Coherent scattering is scattering in which the state of the target is unchanged, except for recoil of the center of mass. The term in (3.27) proportional to $\mathbf{s}_N \cdot \mathbf{F}$ can give scattering in which the neutron and target spin directions change in the scattering. If the nuclear forces were spin independent $\mathbf{s}_N$ and $\mathbf{F}$ would be a constants of the motion and should not change. In this case we would have $a_t = a_s$, and the term would vanish.

For orthohydrogen

$$\sigma_{ortho} = \sigma_{para} + 4\pi (4/3)^2 (a_t - a_s)^2 \sum_{m'M'} |(\mathbf{s}_N \cdot \mathbf{F})_{m'M';mM}|^2,$$

where m and M are the azimuthal spin quantum numbers of

neutron and H_2 in the initial state and m',M' are the corresponding quantities in the final state. On squaring (3.27) the cross term linear in $\underline{s}_N \cdot \underline{F}$ will vanish on averaging over m or M if neutron or target is unpolarized. The last term is easily evaluated, since

$$\sum_{m'M'} |(\underline{s}_N \cdot \underline{F})_{m'M';mM}|^2 = [(\underline{s}_N \cdot \underline{F})^2]_{mM;mM}.$$

Using the familiar relations for spin 1/2 operators and the commutation laws for angular momentum operators we have

$$(\underline{s}_N \cdot \underline{F})^2 = (1/4)\underline{F}^2 + (i/2)\underline{s}_N \cdot (\underline{F} \times \underline{F}) = (1/4)F(F+1) - (1/2)\underline{s}_N \cdot \underline{F}. \tag{3.29}$$

Again the linear term will vanish for unpolarized beam or target, and (3.29), for F=1, equals 1/2. Thus

$$\sigma_{ortho} = \sigma_{para} + 2\pi(4/3)^2(a_t - a_s)^2. \tag{3.30}$$

If we use the values of a_t and a_s given in (2.40) we would expect

$$f_c = 3.719\ f, \quad \sigma_{para} = 3.09\ b, \quad \sigma_{ortho} = 97.9\ b. \tag{3.31}$$

Note the large cancelation between triplet and singlet terms in f_c. As a result of destructive interference between the waves scattered by the two protons the parahydrogen cross section is much smaller than the free N-P cross section, and much smaller than the ortho cross section. By contrast, if a_s had the same sign as a_t (bound singlet state), we should have

$$f_c = -20.0\ f, \quad \sigma_{para} = 89.3\ b, \quad \sigma_{ortho} = 126.7\ b.$$

A measurement of the para cross section would determine the magnitude of f_c. Combining this with the measured epithermal N-P cross section, equations (3.28) and (2.28) can be solved for a_t and a_s. However, it is very difficult to measure the para cross section accurately. The first experiments were carried out, in 1940, by Alvarez and

Pitzer.[a] A later experiment by Sutton et. al.[b] in 1947 gave σ_{para}=4 b, σ_{ortho}=125 b. The results, while not very accurate, clearly showed the destructive interference in parahydrogen scattering. In interpreting the experiments a correction has to be made for the fact that the the neutron wavelength was not very large compared to the H-H separation and a phase difference beween the waves scattered by the two protons has to be allowed for.

Another interesting correction has to do with the fact that the thermal velocities of the H_2 molecules were not negligibly small compared to the neutron velocity. In a medium where the targets are at rest, the reaction rate is R=Nvσ(v), with N the number density of targets. If the targets are moving with a distribution of velocity $\underline{u}$, the rate is R=N$\langle|\underline{v}-\underline{u}|\sigma(|\underline{v}-\underline{u}|)\rangle$, averaged over the distribution of $\underline{u}$. The effective cross section is thus

$$\sigma_{eff}=\langle|\underline{v}-\underline{u}|\sigma(|\underline{v}-\underline{u}|)\rangle/v.$$

For instance, for v very small the effective cross section would be large. In this case collisions are due to the targets bumping the neutrons, rather than vice-versa. The chief source of error in determining σ_{para}, however, results from σ_{para} being so much smaller than σ_{ortho}. A very slight departure from the expected concentration ratios would cause a large error in σ_{para}. More recent measurements by Squires and Stuart[7] in 1955 used H_2 at 20° K and neutrons cooled to liquid oxygen temperature (90° K). The value of the coherent scattering amplitude obtained from measurements of the parahydrogen cross section is f_c=3.80±0.05 f.

[a] L.A. Alvarez and K.S. Pitzer, Phys. Rev., 58, 1003 (1940).
[b] Sutton, Hall, Anderson, Bridge, De Wire, Lavatelli, Long, Snyder and Williams, Phys. Rev., 72, 1147 (1947).
[7] G.L. Squires and A.T. Stewart, Proc. Roy. Soc., A-230, 19 (1955).

SEC.3III DIFFRACTION SCATTERING

Slow neutrons scattered from crystals show the same
diffraction phenomena as do X-rays. The crystal can be
regarded as a large molecule ($\mu_A = M$), and if it is not so
large that absorption effects become significant, the
differential cross section is (see (3.23))

$$d\sigma = |\Sigma_i e^{i\underline{q}\cdot\underline{r}_i} (M/\mu_i) f_i|^2 d\Omega. \qquad (3.32)$$

In the forward direction

$$d\sigma = |\Sigma_i (M/\mu_i) f_i|^2 d\Omega. \qquad (3.33)$$

In a crystal there are values of $\underline{q}$ other than zero for
which the scattered waves add in phase (the Bragg
condition). At the Bragg angle the scattered amplitudes may
again add up to (3.33).

A word should be said about the spin dependence of the
scattering. For a nucleus of spin I, the scattering can be
in states of $I\pm1/2$. The generalization of (3.26) is

$$f_{CM} = -[1/(2I+1)][(I+1)a_{I+1/2} + Ia_{I-1/2} + 2(a_{I+1/2} - a_{I-1/2})\underline{s}\cdot\underline{I}],$$

or, in the laboratory system

$$f = f_c - (M/\mu_i)[2/(2I+1)](a_{I+1/2} - a_{I-1/2})\underline{s}\cdot\underline{I},$$

with

$$f_c = -(M/\mu_i)[1/(2I+1)][(I+1)a_{I+1/2} + Ia_{I-1/2}],$$

which is called the coherent scattering amplitude. Note
that in f_c the amplitudes of the $I\pm1/2$ states are weighted
according to their statistical weights. In the crystal the
directions of the nuclear spins are uncorrelated, and we
may use a representation in which the state is specified by
the azimuthal spin quantum numbers, M_i. If we consider
scattering to all final spin states, the same argument
which led to (3.30) gives in this case, for an unpolarized
crystal

$$d\sigma = \{(\Sigma_i f_{c,i})^2$$
$$+ \Sigma_i [(M/\mu_i)\frac{2}{(2I_i+1)}(a_{i,I_i+1/2} - a_{i,I_i-1/2})]^2 \cdot (1/4) \ I_i(I_i+1)\} d\Omega.$$

No cross terms appear in the last term because the nuclear spin directions are uncorrelated. For a large crystal, composed of N atoms, the first term on the right is proportional to N^2, the second to N. Only the coherent amplitudes are significant in the scattering at the Bragg angles.

At finite temperatures we are not dealing with a perfectly spatially arranged crystal. The normal modes of the crystal are lattice vibrations, and at room temperature many modes are excited. One requires a statistical average of (3.32) over the normal modes. The deformation of the crystal by lattice vibrations means that, at non-zero angles, the amplitudes will no longer add in perfect phase, and there will be a reduction in the scattered intensity at the diffraction peaks. This effect is familiar in X-ray diffraction, and there is a well developed theory for making the necessary corrections. The chief limitation on the accuracy of determinations of f_c by diffraction experiments lies in the uncertainties of this correction.

It should be mentioned that neutron diffraction studies can supply a valuable supplement to X-ray studies. X-rays are scattered primarily by elements of large Z, and hydrogen is virtually undetectable. On the other hand, the neutron coherent scattering from hydrogen is not smaller than that from other nuclei, and the position of hydrogen atoms can be inferred from neutron diffraction. Our description of Bragg scattering has been a little oversimplified. Actually it is the scatterings from the unit cells that add coherently. At a given Bragg angle (scattering from a certain crystal plane) the scattering from the atoms in the unit cell add with definite phase differences. Comparing the scattered intensity from different planes allows determination of the coherent amplitudes of the various constituents.

An experiment to determine the coherent scattering amplitude of hydrogen was carried out in 1948 by Schull,Wollan,Morton and Davidson[*]. Thermal neutrons from a nuclear reactor were monochromatized by Bragg scattering from a rock salt crystal. The powder-crystal method was then used to study the diffraction of these neutrons. The instrument was calibrated by scattering from diamond. Carbon is 99% C^{12}, and, since C^{12} has zero spin, its coherent scattering amplitude is

$$f_c = \pm(13/12)\sqrt{(\sigma/4\pi)}.$$

For C, $\sigma = 4.743 \pm 0.005$ b, and we find $f_c = -6.656 \pm 0.004$ f, assuming the scattering length is positive. The intensity of scattering from other crystals was measured relative to diamond. The coherent scattering amplitude of hydrogen was determined by the diffraction from NaH. The result was $f_c = 3.9 \pm 0.3$ f.

Later experiments reduced the error. The best result for the hydrogen coherent scattering amplitude determined by diffraction is

$$f_c = 3.9 \pm 0.1 \ f.$$

[*] Schull,Wollan,Morton and Davidson, Phys. Rev., 73, 842 (1948).

SEC. 3IV THE OPTICAL THEOREM

A wave passing through a scattering medium can be described as having an index of refraction. If the wavenumber in free space would be p, in the medium it is $p'=\eta p$. The effect is due to interference of the forward scattered waves with the incident wave, and a relation connecting the forward scattering amplitude with the index of refraction (the optical theorem) can be derived by a simple argument given by G.C. Wick. In this argument we assume that the index of refraction description is valid, and calculate the shift in phase produced by a small change in the number density of scatterers in a thin layer of the

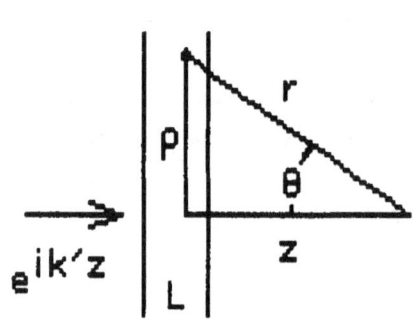

Fig.3.3

medium of thickness 1 (see Fig. 3.3). We take $p'l<<1$, so the phase difference across the layer can be neglected. At the point indicated on the right, the amplitude is changed from $e^{ip'z}$ to $e^{i'z}+\Sigma_i f(\cos\theta_i)e^{ip'r_i}/r_i$ by the additional scatterers, the index i running over the additional scatterers. If we replace the sum by an integral we have

$$e^{ip'z}+dN\int f(\cos\theta)e^{ip'r}dr/r=e^{ip'z}+2\pi l\,dN\int f(\cos\theta)e^{ip'r}\rho d\rho/r,$$

where dN is the change in density of scatterers. Since $r^2=z^2+\rho^2$, we have $rdr=\rho d\rho$ and the integral can be written

$$\int=\int_z^\infty f(\cos\theta)e^{ip'r}dr.$$

We also have $\cos\theta=z/r$, so letting $\zeta=1/\cos\theta=r/z$, we have

$$\int=z\int_1^\infty f(1/\zeta)e^{ip'z\zeta}d\zeta.$$

By partial integration,

$$\int = z[f(1/\mathfrak{z})e^{ip'z\mathfrak{z}}/(ip'z)\,|_1^\infty - (1/ip'z)\int_1^\infty (df(1/\mathfrak{z})/d\mathfrak{z})e^{ip'z\mathfrak{z}}d\mathfrak{z}].$$

A second partial integration shows that the last term on the right is $O(1/p'^2z^2)$. For large $p'z$ we are left with the first term, which gives

$$\int = -(f(1)/ip')e^{ip'z}.$$

There is no contribution from the upper limit because, as we shall see, p' has a positive imaginary part, representing absorption in the medium.

The total wave at z is

$$e^{ip'z}[1+i2\pi l dN f(1)/p']$$
$$=e^{ip'z+i2\pi l dN f(1)/p'}$$

to first order in dN (the higher order terms could be obtained by considering multiple scattering in the layer). The interpretation of the phase shift is that, in the layer of thickness l, p' was changed to p'+dp', with

$$dp' = 2\pi dN f(1)/p'.$$

Thus

$$p'dp' = 2\pi dN f(1),$$

or, integrating,

$$p'^2 - p^2 = 4\pi N f(1), \tag{3.34}$$

since p'=p for N=0. Remember that we wrote f=f(cosθ), so f(1) is the forward scattering amplitude. Since p'$\equiv\eta$p, the formula for the index of refraction is

$$\eta^2 - 1 = 4\pi N f(1)/p^2.$$

This argument indicates that the scattering from distant parts of the medium (p'z>>1) can be described by an index of refraction. The effect of the neighborhood, within a wavelength, depends on the particular circumstances. In general, it can result in the scattering amplitude in the medium differing from the free amplitude. This can happen, for example, if there are spatial correlations in the positions of the scattering centers. A glaring example is the correlation of nucleons within a nucleus. We would not use the sum of the amplitudes of the separate nucleons, but rather the scattering amplitude of the nucleus as a whole. We can invent another simple example of the effect of spatial correlation. Consider the scattering by two fixed

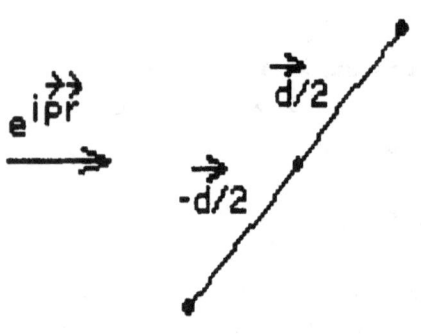

scattering centers separated by a distance $\underline{d}$ (see Fig.3.4). Suppose the scattering length for each center is a, which is large compared to the range of the force, so we can have d small compared to a, but still large compared to the range. The wave scattered by center 2 will have magnitude a/d at 1, and if this is not small

Fig.3.4

compared to one (the magnitude of the incident wave) rescattering effects will be important. We have, for this problem,

$$\underline{\delta}=e^{i\underline{p}\cdot\underline{r}}-a\underline{\delta}_1\frac{e^{ip|\underline{r}-\underline{r}_1|}}{|\underline{r}-\underline{r}_1|}-a\underline{\delta}_2\frac{e^{ip|\underline{r}-\underline{r}_2|}}{|\underline{r}-\underline{r}_2|},$$

with

$$\underline{\delta}_1=e^{i\underline{p}\cdot\underline{r}_1}-a\underline{\delta}_2 e^{ip|\underline{r}_1-\underline{r}_2|}/|\underline{r}_1-\underline{r}_2|,$$
$$\underline{\delta}_2=e^{i\underline{p}\cdot\underline{r}_2}-a\underline{\delta}_1 e^{ip|\underline{r}_2-\underline{r}_1|}/|\underline{r}_2-\underline{r}_1|,$$

and $\underline{r}_1=\underline{d}/2, \underline{r}_2=-\underline{d}/2$. For large r

$$\underline{\delta}\approx e^{i\underline{p}\cdot\underline{r}}-a[\underline{\delta}_1 e^{-i\underline{p}'\cdot\underline{r}_1}+\underline{\delta}_2 e^{-i\underline{p}'\cdot\underline{r}_2}]e^{ipr}/r,$$

where $\underline{p}'$ is the propagation vector of the scattered wave, $\underline{p}'=p\underline{r}/r$. The scattering amplitude is thus

$$f=-a[\underline{\delta}_1 e^{-i\underline{p}'\cdot\underline{d}/2}+\underline{\delta}_2 e^{i\underline{p}'\cdot\underline{d}/2}].$$

The equations for $\underline{\delta}_1$ and $\underline{\delta}_2$ are easily solved, and one finds

$$f=-a[e^{i\underline{q}\cdot\underline{d}/2}+e^{-i\underline{q}\cdot\underline{d}/2}]/[1+(a/d)e^{ipd}]$$
$$=-aN_{\underline{q}}/[1+(a/d)e^{ipd}],$$

with $\underline{q}=\underline{p}-\underline{p}'$ and $N_{\underline{q}}$ the $\underline{q}^{th}$ Fourier component of the density of scatterers. In the long wave length limit, $pd \ll 1$, $f=-2a/(1+a/d)$. For $a/d \gg 1$ we have $f=-2d$, rather than $f=-2a$, which we should have for $a/d \ll 1$. This result is at first sight a little surprising, but we must remember that a scattering length large compared to the range is an unusual condition, meaning that we are close to having a bound state at zero energy. The presence of another scatterer within a wavelength disturbs this condition, and for $a/d \gg 1$ we get a scattering determined by the geometrical size of the scattering system, rather than the abnormally large length a. For scattering of neutrons by nuclei in a medium, d is the interatomic distance and $a/d \approx 10^{-4}$, so rescattering effects are small.

Another derivation of (3.34) can be given for the case of spherically symmetric scattering from a continuous medium, i.e. we imagine $N \to \infty$, Nf remaining finite. We then have

$$\underline{\underline{\delta}}(\underline{r})=\exp(i\underline{p}\cdot\underline{r})+N\int_{vol} f[\exp(ip|\underline{r}-\underline{r}'|)/|\underline{r}-\underline{r}'|]\underline{\underline{\delta}}(\underline{r}')d\underline{r}', \qquad (3.35)$$

where $\underline{r}'$ is the point where the neutron was last scattered and the integral is taken over the volume occupied by scatterer. If we apply the operator $(\Delta+p^2)$ to both sides, we obtain, since $(\Delta+p^2)\exp(i\underline{p}\cdot\underline{r})=0$ and

$$(\Delta+p^2)\exp(ip|\underline{r}-\underline{r}'|)/|\underline{r}-\underline{r}'|=-4\pi\delta(\underline{r}-\underline{r}'),$$
$$(\Delta+p^2)\underline{\underline{\delta}}(\underline{r})=0$$

if $\underline{r}$ is outside the volume, and

$$(\Delta+p^2)\underline{\underline{\delta}}(\underline{r})=-4\pi Nf\underline{\underline{\delta}}(\underline{r})$$

if $\underline{r}$ is inside the volume occupied by the scatterer. Writing the last equation

$$(\Delta+p^2+4\pi Nf)\underline{\underline{\delta}}(\underline{r})=0,$$

we see that in the medium $p'^2=p^2+4\pi Nf$, i.e. in the infinite medium $\underline{\underline{\delta}}(\underline{r})=\exp(i\underline{p}'\cdot\underline{r})$ is a solution.

A physical interpretation of of (3.34) can be given by multiplying by 1/2M. We then have

$$p^2/2M = p'^2/2M - 4\pi Nf(1)/2M = p'^2/2M + \langle V \rangle.$$

This reads as the energy conservation equation, with

$$\langle V \rangle = -4\pi Nf(1)/2M$$

as the mean potential felt by the incident particle in the scattering medium. In fact, a slight adaptation of (3.14) shows us that

$$Nf(1) =$$
$$-(M/2\pi)\int e^{-i\underline{p}\cdot\underline{r}} v_0(\underline{r}_1 \cdots \underline{r}_i \cdots)^* \Sigma_i V_i(\underline{r}-\underline{r}_i)\underline{g}(\underline{r},\underline{r}_1\cdots\underline{r}_i\cdots)$$
$$\cdot d\underline{r}d\underline{r}_1\cdots d\underline{r}_i\cdots/(\text{Volume of scatterer}), \qquad (3.36)$$

where $v_0(\underline{r}_1\cdots\underline{r}_i\cdots)$ is the wave function of the scattering medium. In Born approximation $\underline{g} = \exp(i\underline{p}\cdot\underline{r})v_0$ and

$$\langle V \rangle =$$
$$\int |v_0(\underline{r}_1\cdots\underline{r}_i\cdot)|^2 \Sigma_i V_i(\underline{r}-\underline{r}_i) d\underline{r}d\underline{r}_1\cdots d\underline{r}_i\cdot/(\text{Volume of scatterer}),$$

so $\langle V \rangle$ would be literally the volume average of the potential, averaged over the state of the scattering medium. More precisely, the average potential is a weighted average, weighted, as in (3.36), proportionately to the actual $\underline{g}$. If the medium can exist in a number of states, at any given temperature a statistical average is to be taken over the distribution of states.

The above argument also throws some light on the meaning of $\underline{g}(\underline{r})$ in an expression such as (3.35) when one is not thinking of the limit $n\to\infty, f\to 0$. The true wave function, $\underline{g}(\underline{r},\underline{r}_1\cdots\underline{r}_i\cdots)$ of course varies rapidly when $\underline{r}$ is near any of the $\underline{r}_i$. The smoothed wave function $\underline{g}(\underline{r})$ is obtained by averaging $\underline{g}(\underline{r},\underline{r}_1\cdots\underline{r}_i\cdots)$ over the state v_0:

$$\underline{g}(\underline{r}) = \int v_0(\underline{r}_1\cdots\underline{r}_i\cdots)^* \underline{g}(\underline{r},\underline{r}_1\cdots\underline{r}_i\cdots) d\underline{r}_1\cdots d\underline{r}_i\cdots.$$

Generally, if we go beyond the Born approximation, $f(1)$, and consequently $\langle V \rangle$, has a complex part. This also has a simple interpretation, at least in the case where the index of refraction is close to unity. We can then write (3.34) as $(p'-p)(p'+p) = 4\pi Nf(1)$, and approximately

$$p'-p = 2\pi Nf(1)/p. \qquad (3.37)$$

The magnitude of the wave transmitted through the medium,

with $\underline{p}$ in the z direction let us say, is
$$|\underline{\xi}(\underline{r})|^2 = |e^{ip'z}|^2 = e^{-2Imp'z}.$$

If σ_{tot} is the total cross section for removing particles from the beam, either by elastic scattering or by absorption processes, such as capture of a neutron in the nucleus with gamma emission, the attenuation of the beam should be $\exp(-N\sigma_{tot}z)$. We therefore should have $2Imp' = N\sigma_{tot}$, or, from (3.37),

$$Imf(1)/p = \sigma_{tot}/4\pi. \tag{3.38}$$

The relation (3.38) can easily be checked in the case of scattering by a potential (here $\sigma_{tot} = \sigma_{elastic}$) using the familiar formulae
$$f(\cos\theta) = (1/p)\Sigma_1 (21+1)e^{i\delta_1}\sin\delta_1 P_1(\cos\theta),$$
$$\sigma_{elastic} = (4\pi/p^2)\Sigma_1 (21+1)\sin\delta_1^2.$$
We have, since $P_1(1) = 1$,
$$f(1) = (1/p)\Sigma_1 (21+1)[\cos\delta_1\sin\delta_1 + i\sin\delta_1^2],$$
and (3.38) is seen to be satisfied. It may be noted that (3.38) is satisfied for each 1 separately. This is a consequence of a more general law, of which (3.38) is a special case, called the "unitarity of the S-matrix". The unitarity condition on the scattering matrix is an example of the conservation law for probability, or, less accurately but more graphically, of the conservation of particles in the collision. There is, in fact, a paradox in the usual expression, (2.06), for the wave function at large distances,

$$\xi(\underline{r}) \approx e^{ipr\mu} + f(\mu)e^{ipr}/r, \tag{3.39}$$

with $\mu = \cos\theta$. It appears to say that the incident wave is unchanged, and in addition there is an outgoing scattered wave. How is the number of particles conserved? The answer is that there is destructive interference in the forward direction between the plane wave and the scattered wave which is just sufficient to preserve the conservation law.

To prove this let us calculate the net flux of particles across a large sphere of radius r. The current is given by

$$j=(1/2iM)(\psi^* grad\psi - \psi grad\psi^*),$$

so the radial component of the current is $j_r=(1/M)Im\psi^*\partial\psi/\partial r$ and the flux is

$$F=\int j_r r^2 d\Omega=(2\pi r^2/M)\int_{-1}^{1} Im\psi^*(d\psi/dr)d\mu.$$

If we call the integral I we find from (3.39)

$$I= \int_{-1}^{1} [e^{-ipr\mu}+f(\mu)^*\frac{e^{-ipr}}{r}][ip\mu+ipf(\mu)\frac{e^{ipr}}{r}(1-\frac{1}{ipr})]d\mu$$

$$=\int_{-1}^{1}[ik\mu+\frac{e^{ipr}}{r}ipe^{-ipr\mu}f(\mu)(1-\frac{1}{ipr})+\frac{e^{-ipr}}{r}ip\mu e^{ipr\mu}f(\mu)^*$$

$$+ \frac{ip|f(\mu)|^2}{r^2}]d\mu.$$

We have dropped a term proportional to $(1/r^2)(1/pr)$ since pr can be taken arbitrarily large. The integral of the first term vanishes, the next two can be evaluated to order $1/r^2$ by a partial integration. Integration of $e^{\pm ipr\mu}$ throws down a factor $\pm 1/ipr$, so only the term at the integration limits gives a contribution of order $1/r^2$; the remaining integral can again be partially integrated, which throws down another factor $1/ipr$ and gives terms of order $1/pr^3$. We thus obtain, to order $1/r^2$,

$$I=-\frac{e^{ipr}}{r^2}e^{-ipr\mu}f(\mu)|_{-1}^{1} + \frac{e^{-ipr}}{r^2}e^{ipr\mu}\mu f(\mu)^*|_{-1}^{1}$$

$$+(ip/r^2)\int_{-1}^{1}|f(\mu)|^2 d\mu$$

$$=-(1/r^2)[f(1)-f(-1)e^{2ipr}-f(1)^*-f(-1)^*e^{-2ipr}-ip\int_{-1}^{1}|f|^2d\mu].$$

For F we need the imaginary part of I, which is simply

$$ImI=-[2Imf(1)-p\int_{-1}^{1}|f|^2 d\mu]/r^2,$$

the sum of the two terms involving $f(-1)$ being purely real. We then have

$$F=-(p/M)(4\pi Imf(1)/p -\int|f|^2d\Omega)=-v(4\pi Imf(1)/p -\sigma_{el}),$$

where σ_{el} is the elastic cross section. If there is only elastic scattering, conservation of particles means F=0, which gives us (3.38) for this case. If there is also an absorption cross section, the absorption rate is $v\sigma_{abs}$, and we should have $F=-v\sigma_{abs}$, the minus sign indicating that it is an inward flux. This gives (3.38) with $\sigma_{tot}=\sigma_{el}+\sigma_{abs}$.

There is also a paradox connected with our earlier derivation of the optical theorem for a continuous medium. In this case $\underline{\varrho}=e^{ip'r}$ is an exact solution of (3.35) within the scattering medium. There is no scattering within the medium; scattering only occurs at the surface. This is consistent with the limit N→∞, Nf remaining constant. We have $f=(1/p)e^{i\delta}\sin(\delta)$, and if Nf is to remain finite δ must be proportional to 1/N. For small δ, $f=(\delta+i\delta^2)/p$. In the limit the imaginary part of Nf vanishes, and there is no attenuation of the beam. On the other hand, for a dilute gas of discrete particles one would expect the scattering and absorption to be just the sum of that due to the particles acting separately. The answer is that a discrete gas differs from a continuous medium in one important respect: in the gas there are density fluctuations, and these give rise to scattering.

To show this in a simple case, let us consider the scattering of a particle of mass M by a gas consisting of much heavier particles (taken heavy so the scattering is spherically symmetric in the laboratory system) described by a scattering amplitude f. We shall describe the interaction between incident and target particle by a pseudopotential. This is an algorithm introduced by Fermi, similar in idea to the zero range approximation discussed in Sec. 2IV. The interaction potential, V(r), is taken to be

$$V(\underline{r})=-(2\pi/M)f\delta(\underline{r}).$$

The pseudopotential is intended to be used in first Born

approximation; it is chosen so that the Born approximation amplitude for free scattering, from an initial state of the incident particle of momentum p to a final state of momentum p', equals f:

$$f_{p'p} = -(M/2\pi)V_{p'p} = f\int e^{i(p-p')\cdot r}\delta(r)dr = f.$$

Using the pseudopotential, the Born amplitude for a scattering in which the incident particle goes from state p to state p' and the gas from state n to state n' is

$$f_{n'n} = -(M/2\pi)V_{p'n';pn} = -(M/2\pi)\int e^{i(p-p')\cdot r}[\Sigma_i V_i(r-r_i)]_{n'n}dr$$

$$= f\int \exp(i q\cdot r)[\Sigma_i \delta(r-r_i)]_{n'n}dr.$$

Here $q=p-p'$ is the momentum transfer, and the sum runs over all the particles in the medium. The sum, $\Sigma_i\delta(r-r_i)$, is just the particle density of the medium at r, $N(r)$, so

$$f_{n'n} = f\int \exp(i q\cdot r)N(r)_{n'n}dr.$$

If we were to put $N(r)$=const (continuous medium) we should have $N_{n'n}=0$ for n' not equal to n, and there would be no incoherent scattering. In fact, the incoherent scattering is

$$|f|^2 d\Omega\int \exp(-i q\cdot(r'-r))\Sigma_{n',n'\neq n}N(r')_{nn'}N(r)_{n'n}$$

$$= |f|^2 d\Omega\int \exp(-i q\cdot(r'-r))[(N(r')N(r))_{nn}-N_{nn}^2]dr dr'.$$

The diagonal element, N_{nn}, is the average density in the gas. If we change integration variables from $dr dr'$ to $dr d(r'-r)$ we see that the scattering per unit solid angle per unit volume at momentum transfer q is

$$|f|^2[(N(r')N(r))_{nn}-N_{nn}^2]_q, \qquad (3.40)$$

the q denoting the Fourier component with respect to $r'-r$: the scattering per unit volume with momentum transfer q is proportional the qth Fourier component of the two-particle correlation function. It thus is proportional to density fluctuations occurring in regions of size $1/q$.

The correlation coefficient is easily evaluated for a classical perfect gas. Let us quantize in a box of volume V. The wave function of state n is the product of wave functions $\emptyset_i(r_i)$ of all the scatterers. For our purpose,

the evaluation of $(N(\underline{r}')N(\underline{r}))_{nn}$, we have only to know that
$|\emptyset_i|^2 = 1/V$. We have

$$N(\underline{r}')N(\underline{r}) = \Sigma_{ij}\delta(\underline{r}-\underline{r}_i)\delta(\underline{r}'-\underline{r}_j)$$
$$= \Sigma'_{ij}\delta(\underline{r}-\underline{r}_i)\delta(\underline{r}'-\underline{r}_j) + \Sigma_i\delta(\underline{r}-\underline{r}_i)\delta(\underline{r}'-\underline{r}_i), \qquad (3.41)$$

the prime on Σ' meaning the i=j terms are to be omitted.
In evaluating the diagonal matrix element, each δ function
factor becomes $\int\delta(\underline{r}-\underline{r}_i)|\emptyset_i|^2 d\underline{r}_i = 1/V$. If there are A
scatterers in the gas, the sum with i not equal to j gives
$(A^2-A)/V^2$; in the limit $V\to\infty$, with $A/V = N_{nn}$, this is just
N_{nn}^2, which cancels the last term in (3.40). The remaining
term in (3.41), which can also be written $\Sigma_i\delta(\underline{r}-\underline{r}_i)\delta(\underline{r}'-\underline{r})$
contributes to the diagonal matrix element $(A/V)\delta(\underline{r}'-\underline{r})$.
The Fourier component of the δ function is one, so the
scattering per unit solid angle per unit volume is just
$$N_{nn}|f|^2,$$
as one would expect. If we were considering a Bose or
Fermi gas, the difference from the foregoing calculation
would be that for i not equal j there are, in addition to
the "ordinary" terms we have considered, exchange terms due
to the symmetrization or anti-symmetrization of the wave
function. For example, for a Fermi gas, scatterings with
small momentum transfers will be forbidden if the final
state is already occupied, so one would expect the
scattering per unit volume to be reduced.

Some liquids have unusually large density fluctuations
when near their critical points. The scattered light gives
rise to a striking optical phenomenon known as critical
opalescence.

SEC.3V TOTAL INTERNAL REFLECTION OF NEUTRONS

Optical phenomena familiar for light waves also occur for neutron waves. On passage from a medium of index of refraction n_1 to another of index n_2 there will be a reflected and a refracted wave, as shown in Fig.3.5. The

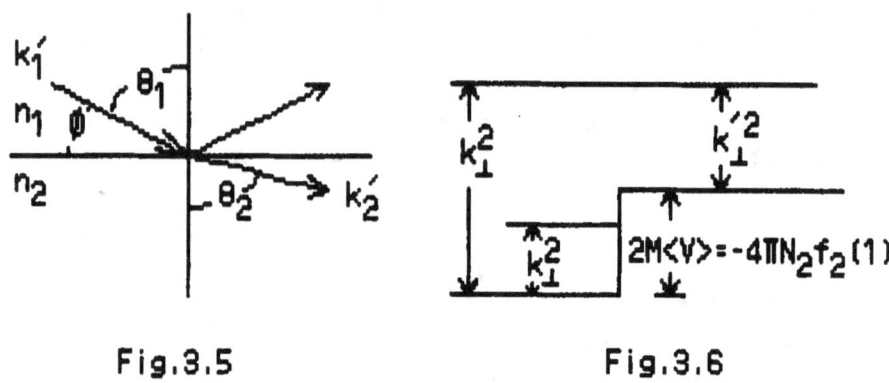

Fig.3.5   Fig.3.6

direction of the refracted wave is given by Snell's Law,
$$\sin\theta_1/\sin\theta_2 = n_2/n_1.$$
If $n_1 > n_2$ then $\theta_2 > \theta_1$, and there is a critical angle of incidence at which $\theta_2 = \pi/2$,
$$\sin\theta_{crit} = n_2/n_1.$$
For $\theta_1 > \theta_{crit}$ there is no transmitted wave and we have total internal reflection.

Since $p_1' = n_1 p$ and $p_2' = n_2 p$, Snell's Law can be written $p_1' \sin\theta_1 = p_2' \sin\theta_2$, or $p_{1||}' = p_{2||}'$, the $||$ sign meaning the component paralell to the surface. On the other hand, from (3.34)
$$p_1'^2 - 4\pi N_1 f_1(1) = p_2'^2 - 4\pi N_2 f_2(1),$$
and, since the parallel components of p_1' and p_2' are equal,
$$p_{1\perp}^2 - 4\pi N_1 f_1(1) = p_{2\perp}^2 - 4\pi N_2 f_2(1). \tag{3.43}$$
Remembering that $-4\pi N f(1) = 2M\langle V \rangle$, we see that (3.43) is the energy conservation equation for a one dimensional problem (motion perpendicular to the surface between the media) in

which the potential energy makes a sudden jump. For slow
neutrons in a solid or liquid medium 2, η_2 is still close
to one; for a gas as medium 1, because N is so much
smaller, we can put $\eta_1=1$. Thus for slow neutrons passing
from air into a solid or liquid we have the picture
illustrated in Fig.3.6. If $f_2(1)$ is negative (positive
scattering length), $\eta_2<1$ and $\langle V_2 \rangle$ is positive. So if
$p_\perp^2>-4\pi N_2 f_2(1)$ we get both a transmitted and reflected wave;
if $p_\perp^2<-4\pi N_2 f_2(1)$ there is only a reflected wave. The latter
is the case of total internal reflection.

To get an idea of the magnitudes involved, let us
consider graphite as the scattering medium. We have seen in
Sec.3III that, for C, $f_c^{(C)}=-6.656$ f. The particle density
is $N=(6.02\cdot10^{23})(2.23)/12=1.13\cdot10^{23}/cm^3$ (the first factor
is Avagadro's number, the second is the density of
graphite, 2.23 gms/cm^3, and 12 is the atomic number). For
room temperature neutrons (E=1/40 ev) $p=0.347\cdot10^8$ cm^{-1}, and
we obtain $4\pi N f_c^{(C)}/p^2=-7.78\cdot10^{-6}$ (i.e. $\eta_2-1=-3.89\cdot10^{-6}$).
The critical angle is thus close to $\pi/2$. If $\emptyset$ is the
complement of θ_1, we have $p_\perp=p\cos\theta_1=p\sin\emptyset\approx p\emptyset$, and we see
from Fig.3.6 that the critical value of $\emptyset$ is
$$\emptyset_{crit}^2=-4\pi N f_c^{(C)}/p^2=7.78\cdot10^{-6},$$
or
$$\emptyset_{crit}=2.79\cdot10^{-3} \text{ radians}=10' \text{ of arc.}$$
Experiments to determine the coherent scattering amplitude,
f_c, by measuring the critical angle must be carried out at
grazing incidence, and were not possible until such strong
thermal neutron beams were available from nuclear reactors
that it was possible to collimate a beam to a fraction of a
milliradian.

The fact that graphite can give total internal
reflection confirms the negative sign of the scattering
amplitude assigned in Sec.3III.

For a hydrocarbon of average composition CH_n,

$$Nf_n(1) = N_C f_C^{(C)} + N_H f_C^{(H)}$$
$$= N_C (f_C^{(C)} + n f_C^{(H)})$$
$$= 6.02 \cdot 10^{23} (\rho/A)(f_C^{(C)} + n f_C^{(H)}), \qquad (3.44)$$

with ρ the density of the hydrocarbon and $A=12+n$. According to (3.31), $f_C^{(H)}$ is positive, so there is destructive interference between the forward scattered waves from C and H. Hydogen itself would not produce total internal reflection, the average potential $\langle V \rangle$ being negative. For benzene, $n=1$, $\rho=0.879$ gm/cm^3, we find $\emptyset_{crit}=1.1 \cdot 10^{-3}=4'$. For paraffin $n=2$ and $Nf_2(1)$ has reversed sign ($n_2>1$) and there is no total internal reflection. If $n=-f_C^{(C)}/f_C^{(H)}$ we should have $Nf_n(1)=0$, $n_2=1$; there is no change in the index of refraction on crossing the surface and the wave is neither reflected nor refracted. Liquids of various n may be used and $f_C^{(H)}/f_C^{(C)}$ determined by measuring the n at which $n_2=1$, $\emptyset_{crit}=0$.

This is the most accurate way to determine f_C. The condition $Nf_n(1)=0$ is independent of p^2, so the energy distribution of the neutrons is irrelevent. For slow neutrons, as we have seen, the coherent forward amplitude is unaffected by form-factors, binding corrections or spatial correlations in the liquid. Surface contamination is unimortant since the wave penetrates far into the liquid; at the critical angle the distance would be infinite if Im p' were truly zero. Actually, the mean free path in benzine is $1/[N_C(\sigma^C+\sigma^H)]=0.25$ cm; the intensity of the wave in the scattering medium falls to 1/e in in this distance. This,however, gives rise to the principle source of error: the background due to neutrons incoherently scattered in the medium.

The experiment was performed by Hughes et. al.[*] in

---

[*] D.J. Hughes, M.T. Burgy and G.R. Ringo, Phys. Rev. 77, 291 (1950); Phys. Rev. 84, 1160 (1951).

1950. Thermal neutrons from the Brookhaven nuclear reactor were collimated by slits 1 mm wide and 6 m apart, giving a beam divergence $\Delta\theta = 1.7 \cdot 10^{-4}$. The intensity of neutrons reflected from the liquid hydrocarbon depends on the reflection coefficient of the surface. For the situation illustratated by Fig. 3.6, the determination of the reflection coefficient, R, is a problem given in every quantum mechanics course. For $p_\perp^2/-4\pi N f(1) < 1$, $R=1$; for $p_\perp^2/-4\pi N f(1) > 1$, $R=(p_\perp - p_\perp')^2/(p_\perp + p_\perp')^2$. The exact form is, however, unimportant for our purpose; the essential feature is that R is a function of $p_\perp^2/-4\pi N f(1) = p^2\theta^2/-4\pi N f(1)$. Thus if the neutrons have an energy spectrum given by $F(p^2)dp^2$ the observed intensity will be

$$I = \int F(p^2) R(p^2\theta^2/-4\pi N f(1)) dp^2$$
$$= I(\theta^2/-4\pi N f(1)). \tag{3.45}$$

The experiment was performed by observing the reflected intensity for one liquid, then, for other mixtures with different n's, determining the angles of incidence, θ_n, which gave the same intensity. Since the intensities were the same, it follows from (3.45) that the values of $\theta_n^2/-4\pi N f_n(1)$ were the same, or $-4\pi N f_n(1) = const \cdot \theta_n^2$. According to (3.44) this means

$$f_c^{(C)} + n f_c^{(H)} = const \cdot (A/\rho)\theta_n^2.$$

Thus if $(A/\rho)\theta_n^2$ is plotted against n a straight line should result; the value of n at which the straight line intersects the horizontal axis gives $f_c^{(C)} + n f_c^{(H)} = 0$; $-f_c^{(C)}/f_c^{(H)} = n$. The best determination is[10] $n = 1.7904 \pm 0.0006$. It is this determination which gave

$$f_c^{(H)} = -3.719 \pm 0.002 \; f$$

---

[10] L. Koester, Z. Phys. <u>198</u>, 187; <u>203</u>, 515 (1967)

and the values of a_t and a_s quoted in Chapter 2.

An interesting variant of the above considerations occurs if medium 2 is a ferromagnet. Then the mean potential in the medium is given by

$$2M\langle V \rangle = -4\pi Nf(1) - 2M\underline{\mu} \cdot \underline{B},$$

where $\underline{\mu}$ is the neutron magnetic moment and $\underline{B}$ is the macroscopic magnetic field (remember that $\underline{B}$ is defined as the volume average of the microscopic field). In terms of μ_N measured in nuclear magnetons, $\underline{\mu} = (e/2M)\mu_N \underline{\sigma}$. If we take for the magnitude of $\underline{B}$ the saturation field $B = 4\pi M$, and express M (the magnetic moment density) in terms of Bohr magnetons per atom, n_B, we have $B = 4\pi N(e/2m)n_B$ with m the electron mass, and

$$2M\langle V \rangle = -4\pi N[f(1) \pm (e^2/2m)\mu_N n_B],$$

the $\pm$ sign referring to neutrons with spins parallel or antiparallel to $\underline{B}$. The value of the neutron moment is $\mu_N = -1.91316$ and e^2/m is the classical electron radius, $e^2/m = 2.8179$ f. For iron $f(1) = -9.2 \pm 0.2$ f and $n_B = 2.219$, and we obtain

$$2M\langle V \rangle = 4\pi N[9.2 \pm 5.98].$$

For cobalt $f(1) = -3.78$ f and $n_B = 1.715$, and

$$2M\langle V \rangle = 4\pi N[3.78 \pm 4.62].$$

Thus in cobalt the parallel polarized beam can show total internal reflection while the antiparallel beam does not; in this way highly polarized neutron beams can be produced by scattering from cobalt mirrors[11].

The circumstance that the two terms in $2M\langle V \rangle$ can be so nearly equal appears to be completely fortuitous. The scale of the first term is set by the range of nuclear forces, that of second by the classical electron radius. Present knowledge gives no clue as to why these two quantities are of the same order of magnitude.

---

[11] D.J. Hughes and M.T. Burgy, Phys. Rev. 75, 1766 (1949)

CHAPTER 4

PHOTOEFFECT OF THE DEUTERON

The photodisintegration reaction

$$\gamma+D\to P+N \tag{4.01}$$

was first observed by Chadwick and Goldhaber[1] in 1934. The threshold energy of this reaction determines the binding energy of the deuteron. Also, since the proton and deuteron masses are known from chemical and mass-spectroscopic measurements, the mass of the neutron is determined as well.

SEC.4I ELECTRIC DIPOLE DISINTEGRATION

A photon of energy 5 Mev has a wavelength $1/k=39.47$ f. The radius of the deuteron is $R=1/\beta=4.316$ f, so $kR=0.11$. For gamma ray energies of this magnitude and smaller we would expect the electric dipole approximation to be good. In this approximation the interaction energy with the electromagnetic field can be taken to be

$$V=-\underline{P}\cdot\underline{E}(0). \tag{4.02}$$

Here $\underline{E}(0)$ is the electric field at the deuteron center of

---

[1] J. Chadwick and M. Goldhaber, Nature <u>134</u>, 237 (1934).

mass, which we take to be at the origin, and P is the electric dipole moment, $P= er_p=(1/2)e(r_P-r_N)$.

An interesting point of physics is buried in (4.02). In general, the absorption of light with propagation vector (k,k) is proportional to $|j_{-k,k}|^2$, i.e. to the square of the four-dimensional Fourier component of $j(r,t)$. The electric dipole approximation neglects retardation and replaces this with $|j_{0,k}|^2$, that is, with the square of the time Fourier component of the volume integral of j. By using the current conservation law the volume integral of j can be expressed in terms of the electric dipole moment: by a partial integration we prove that

$$\int j dr == \int r div j dr,$$

and, from the conservation law $\partial\rho/\partial t+div j=0$ (with ρ the charge density),

$$\int j dr = \int (\partial\rho/\partial t) r dr = dP/dt$$

where $P=\int \rho r dr$ is the electric dipole moment. The time Fourier component $j_{0,k}=(dP/dt)_k=-ikP_k$. The physical question arises in putting $P=er_p$. The N-P interaction is in part an exchange force, due to exchange of charged mesons (Fig.1.4). The contribution of positive and negative mesons add in j, but cancel in ρ. Thus one expects that only the proton charge contributes to P; by taking the interaction in the form (4.02), rather than in terms of $j \cdot A$, one avoids the question of exchange currents.

The cross section for absorption of a photon of momentum k by the deuteron, leading to a state of the continuum with relative momentum p, is

$$d\sigma=2\pi |V_{p;Dk}|^2 p^2 d\Omega/[(2\pi)^3 (dE/dp)] \tag{4.03}$$

The energy equation is

$$k-\epsilon_D=E=p^2/2\mu=p^2/M, \tag{4.04}$$

and $dE/dp=2p/M$. Using (4.02) for V, and writing r for the relative coordinate, $r=r_P-r_N$, we have

$$d\sigma = (e^2 M/32\pi^2) | \int e^{-i\underline{p}\cdot\underline{r}} \underline{r} \cdot (\underline{E}(0))_{0,\underline{k}} \emptyset_D(r) d\underline{r} |^2 p d\Omega, \qquad (4.05)$$

where $\emptyset_D(r)$ is the deuteron wavefunction and $(\underline{E}(0))_{0,\underline{k}}$ is the matrix element of $\underline{E}$ between the one-photon and zero-photon states. The expression for $\underline{E}$ in terms of creation and annihilation operators is

$$\underline{E}(\underline{R}) = i\Sigma_{\underline{k},\lambda} \sqrt{(k/2)} \underline{\epsilon}_{\underline{k},\lambda} [a_{\underline{k},\lambda} e^{i\underline{k}\cdot\underline{R}} - a^+_{\underline{k},\lambda} e^{-i\underline{k}\cdot\underline{R}}].$$

In this form $\underline{E}$ is in Heaviside units and the fine structure constant is $\alpha = e^2/4\pi$. The matrix element for the annihilation of a photon is the coefficient of the annihilation operator, $a_{\underline{k},\lambda}$, so

$$d\sigma = (e^2 M k/64\pi^2) | \int e^{-i\underline{p}\cdot\underline{r}} \underline{r} \cdot \underline{\epsilon}_{\underline{k},\lambda} \emptyset_D(r) d\underline{r} |^2 p d\Omega, \qquad (4.06)$$

with $\underline{\epsilon}_{\underline{k},\lambda}$ a unit vector in the direction of polarization of the photon.

A word must be said about the representation of the final state by a plane wave. The deuteron wave function, $\emptyset_D$, is spherically symmetric and $\underline{r} \cdot \underline{\epsilon}_{\underline{k},\lambda}$ transforms like a first spherical harmonic under rotations of $\underline{r}$. The angular integration in $d\underline{r}$ will thus pick out the p wave part of $e^{-i\underline{p}\cdot\underline{r}}$; the transition is from the deuteron ^{3}S state to a ^{3}P state of the continuum. As we have seen in Sec.2VI on the effective range, the range of nuclear forces is small compared to the deuteron radius, $1/\beta$. The contributions to the integral in (4.06) come from large r, of order $1/\beta$, both because of the factor $\underline{r}$ and because the p wave is small at small r. Using the plane wave amounts to using the free particle p wave, and neglects the p wave phase shift due to nuclear forces. However, this is small for the energies we are considering; it has been neglected in all the scattering calculations.

The integral in (4.06) can be written

$$\int = i\underline{\epsilon}\cdot grad_{\underline{p}} \int e^{-i\underline{p}\cdot\underline{r}} \emptyset_D(r) d\underline{r} = i\underline{\epsilon}\cdot grad_{\underline{p}} \emptyset_{D,\underline{p}}$$

where $\emptyset_{D,\underline{p}}$ is the Fourier coefficient of $\emptyset_D(r)$. The cross section thus depends on the probability the proton has momentum $\underline{p}$ in the deuteron.

In view of the argument that the contributions to the integral come from $r\approx 1/\beta$, it would seem adequate to use the zero range approximation for $\emptyset_D(r)$, i.e. take $\emptyset_D(r)\approx u(r)/r=U(r)/r$. There is, however, a correction necessary for non-zero effective range. In Sec.2VI we used the normalization $u(r)\approx U(r)$, $U(0)=1$. But in (4.06) $\emptyset_D(r)$ must be normalized to unity, $\int|\emptyset_D(r)|^2 d\underline{r}=1$; let us call the radial functions normalized by this condition Nu and $N'U$. For example, $U=e^{-\beta r}$, and, in the zero range approximation the normalization condition is now

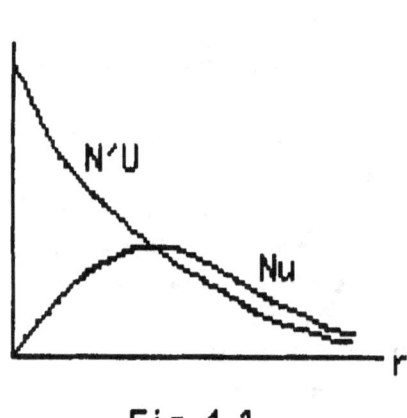

$$4\pi N'^2\int_0^\infty e^{-2\beta r}dr=4\pi N'^2/2\beta=1,$$

or $N'^2=\beta/2\pi$. For finite range, $u(r)<U(r)$ for small r, so it is evident that, to make up for it, Nu must be larger than $N'U$ at large r, i.e. $N>N'$ (see Fig.4.1). We can write

$$\int_0^\infty u^2 dr=\int_0^\infty U^2 dr-\int_0^\infty (U^2-u^2)dr.$$

According to (2.35), the second term equals $r_t/2$. The normalization condition for Nu is thus

Fig.4.1

$$4\pi N^2\int_0^\infty u^2 dr=4\pi N^2[\int_0^\infty U^2 dr-r_t/2]=4\pi N^2[1/2\beta -r_t/2]$$
$$=(N^2/N'^2)(1-\beta r_t)=1.$$

Thus

$$N^2=N'^2/(1-\beta r_t).$$

The correctly normalized wave function at large r is
$$\emptyset_D\approx Ne^{-\beta r}/r,$$
and the Fourier transform of this wave function is
$$\emptyset_{D,\underline{p}}=4\pi N/(p^2+\beta^2),$$
which gives
$$i\underline{\epsilon}\cdot grad_{\underline{p}}\emptyset_{D,\underline{p}}=-i8\pi N\underline{\epsilon}\cdot\underline{p}/(p^2+\beta^2)^2.$$

Inserting this value of the integral in (4.06) gives

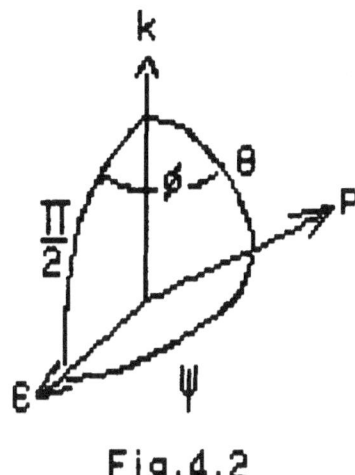

$$d\sigma = \frac{e^2\beta Mk(\underline{\epsilon}\cdot\underline{p})^2 p}{2\pi(1-\beta r_t)(p^2+\beta^2)^4}d\Omega.$$

Since $\epsilon_D=\beta^2/M$, the energy relation, (4.04), can be written

$$Mk=p^2+\beta^2. \qquad (4.07)$$

If ξ is the angle between $\underline{p}$ and $\underline{\epsilon}$, we see from Fig.4.2 that in the spherical triangle

$$\cos\xi=\sin\theta\cos\emptyset.$$

So, using $\alpha=e^2/4\pi$,

$$d\sigma=\frac{2\alpha\beta p^3\sin^2\theta\cos^2\emptyset}{(1-\beta r_t)(p^2+\beta^2)^3}d\Omega.$$

Fig.4.2

For an unpolarized photon beam the average over $\cos^2\emptyset$ gives a factor 1/2 and

$$d\sigma=\frac{\alpha\beta p^3\sin^2\theta}{(1-\beta r_t)(p^2+\beta^2)^3}d\Omega.$$

The p^3 dependence near threshold is due to one power from phase space and two more because the final state is a p state. The total cross section is

$$\sigma_{el}=\frac{8\pi\alpha\beta p^3}{3(1-\beta r_t)(p^2+\beta^2)^3}. \qquad (4.08)$$

We have labeled the cross section σ_{el} to indicate that it is the electric dipole cross section. There is also a magnetic dipole contribution to the photodisintegration cross section which is important near threshold. Comparison with experiment will be postponed until after the magnetic dipole cross section is discussed in the next section.

Eq. (4.08) gives a maximum cross section when $p=\beta$, i.e. when $k=2\epsilon_D$. At the maximum $\sigma=2.40$ millibarns.

Alternatively, since (4.08) is proportional to $1/(1-\beta r_t)$, one may use the measured cross section to determine r_t. The value obtained in this way is $r_t = 1.7 \pm 0.1$ f, in good agreement with that obtained from scattering measurements.

SEC.4II MAGNETIC DIPOLE DISINTEGRATION

Offhand, one might expect the cross section for magnetic dipole photodisintegration of the deuteron to be of order v^2/c^2 compared to the electric dipole cross section. This is a factor of order $\epsilon_D/M \approx 1/400$. This estimate can be misleading for a number of reasons. For one thing, because the nuclear forces are spin dependent, the magnetic dipole transition can go from the deuteron ^3S state to a ^1S state of the continuum. The magnetic dipole cross section near threshold is proportional to the phase space factor p, whereas the electric dipole cross section has an additional factor $(p/\beta)^2$, representing the smallness of the p wave function at the deuteron radius. For small enough p the magnetic cross section is bound to dominate. Also, near threshold the magnitude of the ^1S wave function is enhanced because we are so near to having a bound state at zero energy, i.e. because $-a_s$ is abnormally large. For free particles the s wavefunction would be sinpr/pr=1 at r=0. Actually the s wavefunction is[1] $\exp(-i\delta_s)\sin(pr+\delta_s)/pr$. Near threshold $\delta_s = -pa_s$ and the wavefunction near the origin has the magnitude $1-a_s/r$ rather than 1, as for free particles. At the deuteron radius this factor is $1-\beta a_s = 6.49$, and its square enters the cross section. Finally, there are the anomalous values of the magnetic moments of proton and neutron, $(e/2M)\mu_P$ and $(e/2M)\mu_N$: $\mu_P = 2.79270$ and $\mu_N = -1.91316$. The cross section turns out to have a factor $(1/4)(\mu_P-\mu_N)^2 = 5.5$ not contemplated in the v^2/c^2 estimate.

---

[1] The factor $\exp(-i\delta_s)$, rather than $\exp(i\delta_s)$ as in (2.14) and (2.17), arises because here we choose the boundary condition so that the outgoing rather than the incoming wave in (2.17) agrees with the plane wave (2.16), the appropriate boundary condition for the case when particles are coming out rather than going in.

For the magnetic dipole calculation, $V=-\underline{P}\cdot\underline{E}(0)$ is to be replaced by $V=-\underline{M}\cdot\underline{H}(0)$. In place of (4.06), which can be written

$$d\sigma=(Mk/16\pi^2)\,|\,(\underline{P}\cdot\underline{\epsilon})_{\underline{p},D}|^2pd\Omega,$$

we have

$$d\sigma=(Mk/16\pi^2)\,|\,(\underline{M}\cdot\underline{\epsilon}')_{\underline{p},D}|^2pd\Omega, \qquad (4.09)$$

where $\underline{M}$ is the magnetic dipole moment and $\underline{\epsilon}'$ is a unit vector in the direction of the photon's magnetic field. We shall take $\underline{M}$ to be the sum of the static moments of proton and neutron,

$$\underline{M}=(e/2M)\,(\mu_P\underline{\sigma}_P+\mu_N\underline{\sigma}_N).$$

In this case, unlike the case of the electric dipole, we have no argument that meson exchange currents cannot affect the magnetic dipole moment, other than that the deuteron is a loosely bound system and much of the time the nucleons are outside the range of the forces. We can rewrite the moment

$$\underline{M}=(e/2M)\,[\,(\mu_P+\mu_N)\underline{S}+(1/2)\,(\mu_P-\mu_N)\,(\underline{\sigma}_P-\underline{\sigma}_N)\,], \qquad (4.10)$$

where $\underline{S}=(1/2)\,(\underline{\sigma}_P+\underline{\sigma}_N)$ is the total spin. The matrix element in (4.09) takes the form

$$(\underline{M}\cdot\underline{\epsilon}')_{\underline{p},D}=\int\phi_{\underline{p}}^{(out)}\,(\underline{r})^*\phi_D(r)d\underline{r}(\underline{M}\cdot\underline{\epsilon}')_{S',M';1,M}, \qquad (4.11)$$

the indices S',M' and $1,M$ referring to the S and M_S of final and initial states. The integration over $\underline{r}$ gives zero if $S'=1$, since triplet wave functions of different energies are orthogonal; the final state must have $S'=0$, $M=0$. We shall use the zero-range approximation in this calculation, so $\phi_D=\sqrt{(\beta/2\pi)}e^{-\beta r}/r$. The wavefunction $\phi_{\underline{p}}^{(out)}$ is that for a singlet final state, the superscript "out" indicating that the boundary condition described in Footnote 1 is used, that is, that (at large r) $\phi_{\underline{p}}^{(out)}$ contains a plane wave of momentum $\underline{p}$ and no outgoing spherical wave. The index $\underline{p}$ thus refers to the final momentum of the proton. With the boundary condition used in scattering problems, $\phi_{\underline{p}}^{(in)}$, the

$\underline{p}$ refers to the initial momentum of the scattered particle.
Getting the phase of the matrix element right will not
affect our present calculation but it seemed worth
mentioning; for example it would be necessary if we
considered a polarization experiment in which interference
of electric and magnetic dipole amplitudes were possible.

The angular integration in (4.11) leaves only the
s wave part of $\emptyset_p^{(out)}(\underline{r})^*$, which in zero-range
approximation is $\exp(i\delta_s)\sin(pr+\delta_s)/pr$. The integral in
(4.11) equals

$$4\pi[e^{i\delta_s}/p]J(\beta/2\pi)\int_0^\infty \sin(pr+\delta_s)e^{-\beta r}dr$$

$$=4\pi J(\beta/2\pi)e^{i\delta_s}\sin\delta_s/p)(p\cot\delta_s+\beta)/(p^2+\beta^2). \quad (4.12)$$

Note that if the nuclear forces were spin-independent the
integral would be zero, because the radial wave functions
belonging to different eigenvalues of the energy would be
orthogonal. This appears in (4.12) through the occurrence of
the factor $(p\cot\delta_s+\beta)$. In zero-range approximation
$p\cot\delta_s=-1/a_s$ and $\beta=1/a_t$, so the factor is $[(1/a_t)-(1/a_s)]$.
For spin independent forces we should have $a_t=a_s$, and the
factor would vanish. Note also that as $p\to0$, $\sin\delta_s\to -pa_s$ and
the δ_s dependant terms in the numerator become just $(1-\beta a_s)$,
as expected from the argument given earlier.

More generally, the factor $\exp(i\delta_s)\sin\delta_s/p$ is just the
singlet scattering amplitude (Eq.(2.09)), and thus
(Eq.(2.10))

$$\exp(i\delta_s)\sin\delta_s/p=-1/[(1/a_s)+ip].$$

Eq.(4.12) becomes

$$-\frac{4\pi J(\beta/2\pi)(\beta-1/a_s)}{[(1/a_s)+ip](p^2+\beta^2)}. \quad (4.13)$$

It remains to evaluate the spin part of (4.11). The $\underline{S}$
proportional term of (4.10) will not contribute, since $\underline{S}$
has no matrix elements between singlet and triplet states.

If we take the spin quantization direction in the direction of ϵ', we require the matrix elements of $\sigma_{Pz}-\sigma_{Nz}$. Denoting the spin functions by $\chi_{S,M}$, one readily finds

$$(\sigma_{Pz}-\sigma_{Nz})\chi_{1,\pm 1}=0, \quad (\sigma_{Pz}-\sigma_{Nz})\chi_{1,0}=2\chi_{0,0}.$$

The only non-vanishing matrix element is

$$(\underline{M}\cdot\underline{\epsilon}')_{0,0;1,0}=(e/2M)(\mu_P-\mu_N).$$

Inserting this, (4.13), the energy relation (4.07) and $\alpha=e^2/4\pi$ in (4.09) we find the cross section for the M=0 state to be

$$d\sigma=\frac{\alpha\beta(\beta-1/a_s)^2(\mu_P-\mu_N)^2}{2M^2(p^2+(1/a_s)^2)(p^2+\beta^2)}pd\Omega.$$

For unpolarized deuterons the average cross section is a third of this because only a third of the deuterons are in the M=0 state. The average total cross section is

$$\sigma=\frac{2\pi\alpha\beta(\beta-1/a_s)^2(\mu_P-\mu_N)^2}{3M^2(p^2+(1/a_s)^2)(p^2+\beta^2)}p=\frac{2\pi\alpha}{3}\frac{\epsilon_D}{M}\frac{(1-1/\beta a_s)^2(\mu_P-\mu_N)^2}{(p^2+(1/a_s)^2)(p^2+\beta^2)}\beta p.$$

$$(4.14)$$

At high energies, p>>β, the ratio

$$\frac{\sigma_{mag}}{\sigma_{el}}=\frac{\epsilon_D}{M}(1-1/\beta a_s)^2(1-\beta r_t)\frac{(\mu_P-\mu_N)^2}{4}=0.011.$$

This differs from the offhand estimate primarily by the anomalous moment factor. The maximum of the magnetic dipole cross section occurs near $p^2=1/a_s^2$, i.e. at an energy $1/Ma_s^2=70$ Kev above threshold. As a result of the factors mentioned at the beginning of this section, the magnetic dipole cross section at its maximum is only a factor four smaller than the electric dipole cross section at its maximum.

Fig.4.3.[3] shows the calculated photodisintegration cross section, $\sigma=\sigma_{el}+\sigma_{mag}$, and a number of experimental measurements. The ratio of magnetic to electric cross

[3] L. Hulthen and M. Sugawara, Handbuch der Physik 39, 112, Springer-Verlag, Berlin (1957).

sections can be deduced from observed angular distributions
of the emitted protons.

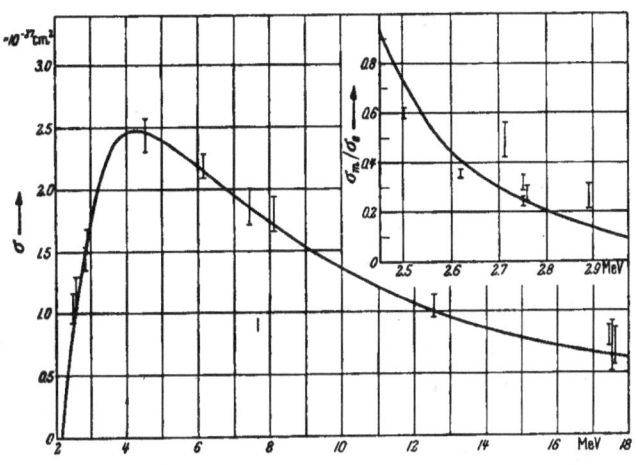

Fig.4.3

SEC.4III THE NEUTRON-PROTON CAPTURE REACTION

The capture reaction,

$$N+P \rightarrow D+\gamma, \qquad (4.15)$$

in which a neutron is captured by a proton with emission of a γ ray, is the inverse of the photodisintegration reaction, (4.01). We can obtain the cross section for (4.15) from that of its inverse by using the argument of detailed balancing. Let us consider the reaction rates $\rho v \sigma$ for an ensemble of the initial and final particles of the reaction in statistical equilibrium in an energy interval ΔE. To maintain the equilibrium the rates of a reaction and its inverse must be equal. Applied to (4.01) an (4.15) this gives

$$2 \cdot 2 \cdot 4\pi p^2 (dp/dE) \Delta E v \sigma_{cap} = 2 \cdot 3 \cdot 4\pi k^2 (dk/dE) \Delta E \sigma_{photo}.$$

The 2·2 on the left is the statistical weight factor for the nucleons, each with s=1/2; the 2·3 on the right is the statistical weight factor for the photon (two polarizations) times that for the deuteron (S=1). Since dp/dE=1/v, dk/dE=1 (remember, in our units c=1) this gives

$$4p^2 \sigma_{cap} = 6k^2 \sigma_{photo}. \qquad (4.16)$$

If we take (4.14) for σ_{photo}, we obtain

$$\sigma_{cap} = \pi\alpha \frac{\epsilon_D}{M} \frac{(1-1/\beta a_s)^2 (\mu_P - \mu_N)^2}{p^2 + (1/a_s)^2} \frac{\beta k}{Mp}. \qquad (4.17)$$

A striking feature of (4.17) is its proportionality to $1/p$, or $1/v$, as $p \rightarrow 0$. This is a common feature of exothermic reactions. For an exothermic reaction the phase space factor for the final particles remains finite as the initial energy goes to zero. We are left with the $1/v$ factor which normalizes the initial current to unit flux. Thus the cross section is proportional to $1/v$, provided the

square of the matrix element does not vanish for $v=0$. For
example, for the electric dipole capture the square of the
matrix element is proportional to p^2; including the
electric dipole cross section on the right in
(4.16) would give a term proportional to p at threshold,
rather than $1/p$. No contradiction is involved in a cross
section which becomes infinite as $v \to 0$. Consider a neutron
traversing a hydrogenous material, with hydrogen density N.
The lifetime of the neutron in the material due to the
capture in hydrogen is $\tau = 1/Nv\sigma_{cap}$. Thus a capture cross
section proportional to $1/v$ simply means that τ is a
constant independent of velocity.

Near zero energy (4.17) becomes

$$\sigma_{cap} = 2\pi a_s^2 \alpha (\epsilon_D/M)^2 (1-1/\beta a_s)^2 (\mu_P - \mu_N)^2 \beta/p_{lab}. \qquad (4.18)$$

where $p_{lab} = 2p$ is the neutron momentum in the laboratory
system. For room temperature thermal neutrons $E_{lab} = 1/40$ ev,
$p_{lab} = 3.47 \cdot 10^8$ cm^{-1}, (4.18) gives

$$\sigma_{cap} = 0.301 \text{ b}.$$

The experimental value is

$$\sigma_{cap} = 0.3347 \pm 0.0008 \text{ b}, \qquad (4.19)$$

11% larger. One's first thought is that use of the zero-
range approximation could easily lead to such a difference.
However, calculations including effective range corrections
have failed to remove the discrepancy. The conclusion is
that the discrepancy gives us a measure of the error
involved in representing the magnetic moment simply in
terms of the static moments of the proton and neutron.

CHAPTER 5

THE SLOWING AND DIFFUSION
OF NEUTRONS

The experiments described in Chapt.3, as also many of the early experiments of Fermi and his group in Rome, depended largely on the technique of slowing neutrons to thermal energies in materials such as paraffin. In this chapter we discuss the slowing down process.

If the elastic scattering of a neutron by a nucleus of mass $M_A = AM$ is spherically symmetrical in the center of mass system, in the laboratory system the cross section for scattering from an initial neutron momentum p' to a final momentum p is proportional to pdp, as we see from (3.05) or (3.11). Thus

$$d\sigma = Kpdp = K'dE.$$

Since $p_{max} = p'$, $p_{min} = [(M_A - M)/M_A + M)]p'$, the energy distribution of scattered neutrons is uniform between the initial energy, E', and the and the minimum energy, $\int E'$, with

$$\int = (M_A - M)^2/M_A + M)^2 = (A-1)^2/(A+1)^2.$$

The total cross section is

$$\sigma = K' \int_{\int E'}^{E'} dE = K'(1-\int)E',$$

so

$$d\sigma=\sigma dE/(1-\mathfrak{f})E', \qquad \mathfrak{f}E'<E<E'. \qquad (5.01)$$

For hydrogen, $\mathfrak{f}=0$, and the energy distribution is
uniform from E' to zero. On the average the neutron loses
half its energy in course of a scattering. For carbon
$\mathfrak{f}=(11/13)^2=0.716$, and, on the average, the neutron loses
only $(1-\mathfrak{f})/2\approx1/7^{th}$ of its energy per collision. For this
reason, and also because at low energy the hydrogen
scattering cross section is larger than that of carbon,
most of the slowing down in a hydrocarbon is due to
hydrogen. So let us forget energy loss due to carbon and
consider the slowing as due solely to hydrogen.

SEC.5I SLOWING IN HYDROGEN

Let us suppose there is a source producing $S(E,t)$ neutrons/sec/unit energy, and let the density of neutrons per unit energy be $\rho(E,t)$. With the scattering cross section given by (5.01) with $\ell=0$, the equation determining ρ is

$$\partial\rho/\partial t = -R\rho + \int_E^\infty R'\rho'dE'/E' + S(E,t) \qquad (5.02)$$

Here R is the collision rate, $R=Nv\sigma$ with N the number density of hydrogen atoms, and a prime indicates the quantity is to be evaluated at E'. We neglect the capture cross section, which is less than $1/400^{th}$ the scattering cross section for energies above 1 ev.

If the source produces monoenergetic neutrons of energy E_0, so

$$S(E,t)=S(t)\delta(E-E_0),$$

ρ takes the form

$$\rho(E,t)=n_0(t)\delta(E-E_0)+n(E,t),$$

where $n(E,t)=0$ for $E>E_0$. Substitution in (5.02) gives the equations

$$\partial n_0/\partial t = -R_0 + S,$$

$$\partial n/\partial t = -Rn + \int_E^{E_0+0} R'[n_0\delta(E'-E_0)+n']dE'/E'$$

$$= -Rn + \int_E^{E_0} R'n'dE'/E' + R_0n_0/E_0. \qquad (5.03)$$

If we write

$$s_0=R_0n_0, \quad s=Rn,$$

the equations (5.03) become

$$\frac{1}{R_0}\frac{\partial s_0}{\partial t} = -s_0 + S,$$

$$\frac{1}{R}\frac{\partial s}{\partial t} = -s + \int_E^{E_0} s'\frac{dE'}{E'} + \frac{s_0}{E_0}. \qquad (5.04)$$

Let us first consider the static case, where $S(t)$ is constant in time, as are also s_0 and s. We then have

$$s_0 = S,$$

$$s = \int_E^{E_0} s' \frac{dE'}{E'} + \frac{S}{E_0} . \qquad (5.05)$$

Differentiating with respect to E we find

$$ds/dE = -s/E,$$

so $s = C/E$. From (5.05), $s(E_0) = S/E_0$; therefore $C = S$ and

$$s = S/E. \qquad (5.06)$$

Fermi introduced the concept of the "slowing down density", $q(E,t)$, which is the number of neutrons falling below energy E per second. For example, if one wanted to know the number of neutrons becoming thermal per second this would be the quantity of interest. With a cross section $d\sigma = \sigma dE/E'$, it is evident that if $A dE$ neutrons/sec fall into dE, the number, q, falling below E is $q = A \int_0^E dE = AE$. The quantity on the right in (5.05) is "A", or q/E, and (5.05) tells us that this equals s. Thus $q = sE = S$. This is an obvious conservation law for the problem with S independent of t and no absorption. More generally, since by definition q is the downward neutron flux past energy E, we have the differential conservation law, for $E < E_0$,

$$\partial n/\partial t = \partial q/\partial E, \qquad (5.07)$$

or, more generally still,

$$\partial \rho/\partial t = \partial q/\partial E + S(E,t). \qquad (5.08)$$

Now let us consider the problem of a neutron of energy E_0 introduced at time $t=0$, so $S(t) = \delta(t)$. The solution of the first of the equation (5.04) is

$$s_0 = 0, \qquad\qquad t < 0,$$

$$s_0 = R_0 e^{-R_0 t}, \quad t > 0 \qquad (5.09)$$

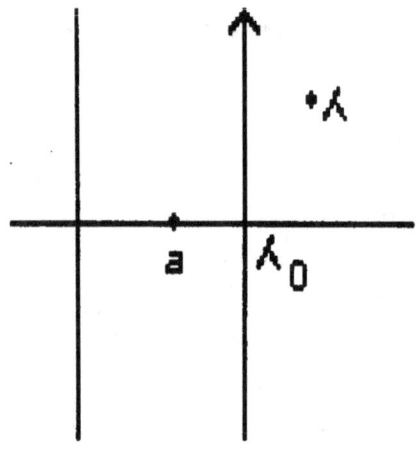

Fig.5.1

Since s is zero for negative t, we shall seek a solution of the second equation in terms of a Laplace integral. This is a transformation in which one expresses a function f(t) (we are concerned with a real function of a real variable) for t>0 in the form

$$f(t)=\frac{1}{2\pi i}\int_{\lambda_0-i\infty}^{\lambda_0+i\infty} F(\lambda')e^{\lambda' t}d\lambda', \quad (5.10)$$

the contour of integration running parallel to the imaginary axis (Fig.5.1). Multiply both sides of (5.10) by by exp(-λt) and integrate with respect to t,

$$\int_0^\infty f(t)e^{-\lambda t}dt=\frac{1}{2\pi i}\int_{\lambda_0-i\infty}^{\lambda_0+i\infty}d\lambda' F(\lambda')\int_0^\infty e^{(\lambda'-\lambda)t}dt.$$

If Re λ>Re λ' the t integral on the right converges and gives -1/(λ'-λ), so

$$\int_0^\infty f(t)e^{-\lambda t}dt=-\frac{1}{2\pi i}\int_{\lambda_0-i\infty}^{\lambda_0+i\infty}\frac{F(\lambda')}{(\lambda'-\lambda)}d\lambda'.$$

If F(λ') is analytic to the right of the contour and →0 as λ'→∞ we can close the contour to the right, and the integral is just -2πi times the residue at the pole, so we obtain the formula for the Laplace inverse

$$F(\lambda)=\int_0^\infty f(t)e^{-\lambda t}dt. \quad (5.11)$$

If f(t) is bounded at the origin and behaves like e^{at} for large t, we see from (5.11) that the conditions on F(λ) will be satisfied if Re λ>a, and our procedure is formally correct if the contour of (5.10) passes to the right of a,

$\lambda_0 > a$.

Note that if t is negative the contour of (5.10) can be closed to the right and, since $F(\lambda')$ is analytic to the right of the contour, the integral is zero, so (5.10) gives

$$f(t)=0, \quad t<0.$$

It is this feature which suggests using the Laplace transform for the present problem.

We therefore write

$$s(E,t)=\frac{1}{2\pi i}\int_{-i\infty}^{i\infty} s_\lambda(E)e^{\lambda t}d\lambda, \tag{5.12}$$

and a similar expression for s_0. These expressions substituted in (5.04) give

$$(1+\lambda/R)s_\lambda=\int_E^{E_0} s_\lambda' dE'/E' + s_{0\lambda}/E_0, \tag{5.13}$$

and, from (5.09) and (5.11)

$$s_{0\lambda}=1/(1+\lambda/R_0). \tag{5.14}$$

If we write

$$w_\lambda=(1+\lambda/R)s_\lambda \tag{5.15}$$

equation (5.13) becomes

$$w_\lambda=\int_E^{E_0} w_\lambda' dE'/(1+\lambda/R')E' + s_{0\lambda}/E_0. \tag{5.16}$$

Differentiating with respect to E we get

$$dw_\lambda/dE=-w_\lambda/(1+\lambda/R)E. \tag{5.17}$$

Let us suppose that over the range of energies of interest R is proportional to $E^{1/k}$, so $dE/E = kdR/R$, with $k>0$. Equation (5.17) becomes

$$dw_\lambda/w_\lambda=-kdR/(R+\lambda),$$

whence

$$w_\lambda=(s_{0\lambda}/E_0)(R_0+\lambda)^k(R+\lambda)^{-k},$$

the constant of integration being chosen so (5.16) is satisfied when $E=E_0$. Using (5.14) and (5.15),

$$s_\lambda=(R_0R/E_0)(R_0+\lambda)^{k-1}(R+\lambda)^{-k-1}. \tag{5.18}$$

The integral resulting from substituting this is in (5.12) can be put in more recognizable form by changing the variable of integration from λ to $u=-(R+\lambda)t$; (5.12) can then be written in the form

$$s(E,t)=(R_0/E_0)Rte^{-Rt} z^{k-1}F_k(z)/\Gamma(k+1), \qquad (5.19)$$

with $z=(R_0-R)t$ and

$$F_k(z)=\frac{\Gamma(k+1)}{2\pi i} \int_{-i\infty}^{i\infty}(-u)^{-k-1}(1-\frac{u}{z})^{k-1}e^{-u}du. \qquad (5.20)$$

For $t>0$, the contour of (5.20) passes to the left of the singularity at the origin and the contour can be deformed to a contour, C, starting at infinity, circling the origin in a clockwise direction, and returning to infinity. The integral in (5.20) is a standard representation of a confluent hypergeometric function.[1] Its behavior for large z is readily found by expanding the integrand as a power series in $1/z$ and integrating term by term by means of Hankel's formula[2]

$$1/\Gamma(x)=(1/2\pi i)\int_C (-u)^{-x}e^{-u}du.$$

Thus

$$F(z)\approx 1+k(k-1)(1/z)+\cdots.$$

If k is a positive integer the series terminates with the kth term and $s(E,t)$ is an elementary function.

We have seen that the slowing down density is E times the last two terms in (5.03), so

$$q(E,T)=E[s+(1/R)\partial s/\partial t]. \qquad (5.21)$$

The Laplace transform of $q(E,t)$ is thus

$$q_\lambda=E(1+\lambda/R)s_\lambda=Ew_\lambda, \qquad (5.22)$$

and $q(E,t)$ can also be expressed directly in a form similar to (5.19) and (5.20).

---

[1] See E.T. Whittaker and G.N. Watson, Modern Analysis, Fourth Edition, Cambridge (1927), p.339.
[2] Whittaker and Watson, loc. cit. p.245.

For large z the leading term of (5.19) gives for the neutron density

$$n=s/R\approx(R_0^k/E_0)t^k(1-R/R_0)^{k-1}e^{-Rt}/\Gamma(k+1), \qquad (5.23)$$

and for q

$$q\approx(R_0^k/E_0)Et^{k-1}(1-R/R_0)^{k-1}e^{-Rt}/\Gamma(k). \qquad (5.24)$$

It is notable that, since it follows from the definition of k that R_0^k/E_0 is a constant independent of E_0, (5.23) and (5.24) approach limits independent of E_0 as $E_0\rightarrow\infty$. The neutron density is proportional simply to t^ke^{-Rt}. For fixed t it is a monotonically decreasing function of E, while at fixed E it has a maximum at $t=k/R$, i.e. at k times the final collision time.

The exact solutions for s and q (as well as the limiting cases (5.23) and (5.24)) satisfy

$$\int_0^\infty s(E,t)dt=1/E, \quad \int_0^\infty q(E,t)dt=1.$$

Equation (5.11) shows that the time integral of a function is just its Laplace transform for $\lambda=0$, so these results follow from $w_{\lambda=0}=s_{\lambda=0}=(R_0^k/E_0) R^{-k}=1/E$. That the time integral of q is one is an obvious conservation law ; the relationship for s is necessary if the time dependent solution is to reduce to the static solution in the case of a static source. For an arbitrary S(t) the solution (call it $s_1(E,t)$ to distinguish it from s(E,t)) is

$$s_1(E,t)=\int_{-\infty}^t s(E,t-t')S(t')dt',$$

which, for constant S, becomes

$$s(E)=S\int_{-\infty}^t s(E,t-t')dt'=S\int_0^\infty s(R,\tau)d\tau. \qquad (5.25)$$

Comparing with the static solution, (5.06), gives us the condition quoted.

If we are interested in the slowing down to thermal or epithermal velocities, it is evident that nearly all the time is spent and nearly all the collisions take place at energies so low that σ remains near its epithermal value. For constant σ, R is proportional to v, so k=2 and the solutions are very simple:

$$n=(1/2)(R_0/E_0)(z+2)te^{-Rt},$$ (5.26)

$$q=((R_0/E_0)(z+1)Ee^{-Rt} .$$

It is easily checked that sensible answers are also given in the z→0 limit. For example, $q(E_0,t)=R_0e^{-Rt}s_0$, as given by (5.09), which is the correct answer. And for $R_0t\ll1$, $q=R_0E/E_0$ and $n=R_0t/E_0$, the densities resulting from the first collision.

In the above calculations the capture cross section has been neglected. It is a simple matter to include the effect of a 1/v capture cross section. Since the capture rate, $R_c=Nv\sigma_{cap}$ is independent of E, the probability of a neutron's surviving for time t is just e^{-R_ct}, and one has only to multiply the solution of the time dependent problem, (5.19), by this factor. For the problem of the energy distribution produced by a constant source, the integral in (5.25) is replaced by

$$\int_0^{\infty}s(E,\tau)e^{-R_c\tau}d\tau,$$

so $s(E)=Ss_{\lambda=R_c}$. Similarly, $q(E)=ESw_{\lambda=R_c}$. In place of $s(E)=S/E$ and $q(E)=S$ we now have

$$s(E)=S(R_0R/E_0)(R_0+R_c)^{k-1}(R+R_c)^{-k-1},$$

$$q(E)=S(R_0E/E_0)(R_0+R_c)^{k-1}(R+R_c)^{-k}.$$

As a check, we observe that $q(E_0)=SR_0/R_0+R_c)$, the fraction of the source that survives the first collision. The limiting forms for $E_0\to\infty$ are

$$s(E)=(S/E)/(1+R_c/R)^{k+1},$$

$$q(E)=S/(1+R_c/R)^k.$$

For slowing to epithermal energy, k=2 and $q=S/(1+\sigma_c/\sigma)^2$. The loss in slowing to 0.1 ev, where $\sigma_c=0.165$ b and $\sigma=20.4$ b, is 1.6%.

The loss in slowing neutrons in hydrogen was not of much significance when one was slowing neutrons from a radioactive source in a block of paraffin. However, in a nuclear reactor the slowing loss, plus the loss of thermal neutrons by capture in hydrogen, make hydrogen a non-too-good neutron moderator. Deuterium and carbon have much smaller low energy capture cross sections, and graphite was the material Fermi used in his first reactor. There is therefore a considerable interest in studying the slowing of neutrons by heavier nuclei.

SEC.5II SLOWING BY HEAVIER NUCLEI

For heavier nuclei we must use (5.01) with $\mathcal{J}$ not equal zero. The second equation of (5.04) then becomes

$$(1/R)\,\partial s/\partial t = -s + (1-\mathcal{J})^{-1}\int_E^{E/\mathcal{J}}[s_0\delta(E'-E)+s']dE'/E'. \qquad (5.27)$$

For a source constant in time we have $s_0 = S$ and

$$s = (1-\mathcal{J})^{-1}\int_E^{E/\mathcal{J}}[S\delta(E'-E)+s']dE'/E'. \qquad (5.28)$$

We require a solution which has $s=0$ for $E>E_0$. If we change variable to

$$y = E_0/E,$$

(5.28) becomes

$$s = (1-\mathcal{J})^{-1}\int_{\mathcal{J}y}^{y}[(S/E_0)\delta(y'-1)+s']dy'/y', \qquad (5.29)$$

with the condition $s(y)=0$, $y<1$.

We can adapt the Laplace transform from the interval $(0,\infty)$ to the interval $(1,\infty)$ by the transformation $t=\ln y$. Equations (5.10) and (5.11) then take the form

$$f(y) = (1/2\pi i)\int_{\lambda_0-i\infty}^{\lambda_0+i\infty} F(\lambda)y^\lambda\,d\lambda,$$

$$F(\lambda) = \int_1^\infty f(y)y^{-\lambda-1}dy.$$

This is known as the Mellin transform. Accordingly, let us write

$$s(y) = (1/2\pi i)\int_{\lambda_0-i\infty}^{\lambda_0+i\infty} s_\lambda y^\lambda d\lambda. \qquad (5.30)$$

Since the Mellin transform of of $\delta(y-1)$ is unity, (5.29) becomes

$$s(y) = \frac{1}{2\pi i}\;\frac{1}{(1-\mathcal{J})}\int_{\lambda_0-i\infty}^{\lambda_0+i\infty}\left(\frac{S}{E_0}+s_\lambda\right)\int_{\mathcal{J}y}^{y}y'^{\lambda-1}dy'\,d\lambda$$

$$=\frac{1}{2\pi i}\ \frac{1}{(1-\mathcal{I})}\int_{\lambda_0-i\infty}^{\lambda_0+i\infty}(\frac{S}{E_0}+s_\lambda)\frac{(1-\mathcal{I}^\lambda)}{\lambda}y^\lambda d\lambda.$$

Comparing with (5.30) gives

$$s_\lambda=\frac{1}{(1-\mathcal{I})}(\frac{S}{E_0}+s_\lambda)\frac{(1-\mathcal{I}^\lambda)}{\lambda},$$

or, solving for s_λ,

$$s_\lambda=\frac{1}{(1-\mathcal{I})}\ \frac{S}{E_0}\ (1-\mathcal{I}^\lambda)[\lambda-\frac{(1-\mathcal{I}^\lambda)}{(1-\mathcal{I})}]^{-1}.$$

We then have

$$s(y)=\frac{1}{2\pi i}\ \frac{1}{(1-\mathcal{I})}\ \frac{S}{E_0}\int_{\lambda_0-i\infty}^{\lambda_0+i\infty}(1-\mathcal{I}^\lambda)y^\lambda[\lambda-\frac{(1-\mathcal{I}^\lambda)}{(1-\mathcal{I})}]^{-1}d\lambda. \qquad (5.31)$$

The contour is to run to the right of the poles of the integrand, which requires $\lambda_0>1$.

For y near one, s(y) can be evaluated by expanding the denominator of (5.31) in powers of $(1-\mathcal{I}^\lambda)/[\lambda(1-\mathcal{I})]$. The nth term in the series gives the contribution of neutrons which have made n collisions. For large y, s(y) can be expressed as a series of residues from the poles of the integrand. The leading term comes from the pole at $\lambda=1$. To calculate the residue we note that near $\lambda=1$
$[\lambda-(1-\mathcal{I}^\lambda)/(1-\mathcal{I})]=d/d\lambda[\qquad]_{\lambda=1}(\lambda-1)=\{1-[\mathcal{I}/(1-\mathcal{I})]\ln(1/\mathcal{I})\}(\lambda-1)$
so the residue gives

$$s(y)\approx Sy/\epsilon E_0=S/\epsilon E, \qquad (5.32)$$

with

$$\epsilon=1-\frac{\mathcal{I}}{1-\mathcal{I}}\ \ln(1/\mathcal{I}). \qquad (5.33)$$

The remaining poles lie in the negative half plane and give contributions, proportional to negative powers of y, which are small for large y. These terms are oscillatory, and give the residual effect of the sharp corners in the distribution near y=1. For example, for $\mathcal{I}=(11/13)^2$, the value for carbon, the nearest poles are at $\lambda=-5.767\pm i22.327$ and give contributions proportional to $y^{-5.767}$.

The leading term, (5.32), is so simple that one might seek a simpler derivation of it. For example, one might try to solve (5.28) by differentiation, as was done in the hydrogen case. If $E < \int E_0$, so the δ-function doesn't contribute, we find

$$ds/dE = [1/(1-\int)][s(E/\int) - s(E)]/E,$$

a functional equation which is solved by

$$s = C/E. \tag{5.34}$$

However (5.28), with the boundary condition $s=0$ for $E > E_0$, is not satisfied for $E > \int E_0$. Thus (5.34) is an exact solution of the problem only for a source at infinite energy.

The constant C can be determined by the condition that the slowing down density, q, equals the source, S. The rate of slowing below energy E is

$$q = \frac{1}{1-\int} \int_E^{E/\int} s' \frac{(E - \int E')}{E'} \, dE'$$

$$= \frac{C}{1-\int} \int_E^{E/\int} \frac{(E - \int E')}{E'^2} dE' = \frac{C}{1-\int} [1 - \int - \int \ln(1/\int)] = C\epsilon = S.$$

Thus $C = S/\epsilon$, in agreement with (5.32).

The above calculation shows that, in the static case,

$$q = \epsilon E s. \tag{5.35}$$

In the time dependent case we have (5.07), which can be written

$$(1/R) \, \partial s/\partial t = \partial q/\partial E. \tag{5.36}$$

If we supposed (5.35) also held in the time dependent case, we should have

$$\partial q/\partial t = \epsilon R E \partial q/\partial E = -\partial q/\partial T,$$

where

$$T = (1/\epsilon) \int_E^{E_0} dE'/R'E', \tag{5.37}$$

an equation which is solved by

$$q = f(t-T).$$

For example, for a neutron of energy E_0 introduced at t=0,
$$q=\delta(t-T).$$

This is an approximation which was introduced by Fermi, who pointed out that the value of $\ln(E/E')$, averaged over a collision, is

$$[1/(1-f)]\int_{fE'}^{E'}\ln(E/E')dE/E'=-\epsilon.$$

Thus $\ln E$ is reduced, on the average, by ϵ per collision, and the average number of collisions in reaching energy E is

$$N_c=(1/\epsilon)\ln(E_0/E). \qquad (5.38)$$

Since 1/R is the time per collision, the time required to slow from E_0 to E is

$$T=\int_0^{N_c}dN_c'/R'=(1/\epsilon)\int_E^{E_0}dE'/R'E',$$

in agreement with (5.37).

Equation (5.35), which we have used in the above argument, is not really correct for the time dependent problem. It is, however, the leading term in a systematic expansion in the case of slowing by a heavy nucleus, where f is close to one, and ϵ, in consequence, is close to zero.

Let us write
$$w=Es \quad \text{and} \quad u=-\ln E.$$
The neutron density on a logarithmic scale is $n(u)=En(E)$, so
$$w=Rn(u).$$
Changing variables in (5.27) gives, for $E<fE_0$,

$$(1/R)\partial w/\partial t=-w+[1/(1-f)]\int_0^{\ln(1/f)}w(u-x)e^{-x}dx, \qquad (5.39)$$

where $x=u-u'$.

If $f=1-\delta$, with $\delta\ll1$, then, to first order in δ, $\epsilon=(1/2)\delta$ and $\ln(1/f)=\delta=2\epsilon$ In the limit $\epsilon\rightarrow0$ it requires many collisions to reduce the energy from E_0 to E, and we

may suppose the width in u of the w distribution is large
compared to 2ϵ, the spread in a single collision. We
therefore expect that (5.39) can be evaluated by expanding
$w(u-x)$ as a Taylor series in x. The first three terms in
the series are

$$\frac{1}{1-\mathfrak{f}}\int_0^{\ln(1/\mathfrak{f})} w(u-x)e^{-x}dx = \frac{1}{1-\mathfrak{f}}\int_0^{\ln(1/\mathfrak{f})}[w(u)-\frac{\partial w(u)}{\partial u}x+\frac{1}{2}\frac{\partial^2 w(u)}{\partial u^2}x^2]e^{-x}dx$$

$$=w(u)-\epsilon\partial w(u)/\partial u+(2/3)\epsilon^2\partial^2 w(u)/\partial u^2.$$

We have neglected terms of order ϵ^3 in the coefficient of
the second derivative term. To order ϵ^2 this coefficient is
just half of

$$<x^2>=(1/2\epsilon)\int_0^{2\epsilon}x^2 dx. \tag{5.40}$$

Equation (5.39) becomes

$$(1/R)\partial w/\partial t=-\epsilon\partial w/\partial u+(2/3)\epsilon^2\partial^2 w/\partial u^2. \tag{5.41}$$

In terms of w and u, the conservation equation,
(5.07), reads

$$(1/R)\partial w/\partial t=-\partial q/\partial u.$$

Comparing this with (5.41) we see that

$$q=\epsilon[w-(2/3)\epsilon\partial w/\partial u]. \tag{5.42}$$

The problem of a neutron of energy E_0 inserted at $t=0$
can be solved by expressing w as a Laplace integral

$$w(u,t)=(1/2\pi i)\int w_\lambda(u)e^{\lambda t}d\lambda. \tag{5.43}$$

Equation (5.41) then gives the expression for the Laplace
transform

$$(2/3)\epsilon^2 d^2 w_\lambda/du^2-\epsilon dw_\lambda/du-(\lambda/R)w_\lambda=0. \tag{5.44}$$

This equation can be solved by an approximation similar to
the W.K.B. approximation for the Schrodinger equation. We
write

$$w_\lambda=\exp(\epsilon^{-1}S_\lambda),$$

which, substituted in (5.44) gives the Ricatti equation

$$(2/3)(dS_\lambda/du)^2-dS_\lambda/du-(\lambda/R)=-\epsilon d^2 S_\lambda/du^2. \tag{5.45}$$

To lowest order in ϵ, dS_λ/du satisfies the quadratic equation obtained by neglecting the ϵ proportional term on the right, of which a root is

$$dS_\lambda/du=(3/4)[1-(1+8\lambda/3R)^{1/2}],$$

giving us

$$S_\lambda=(3/4)\int_{u_0}^{u}[1-(1+8\lambda/3R')^{1/2}]du'+const. \qquad (5.46)$$

In the present case, in contrast to the case of slowing in hydrogen, we may expect the time it takes the neutron pulse to pass the point u to be large compared to 1/R, the time of a single collision. If the width of the pulse in u is called Δu, it will require $\Delta u/\epsilon$ collisions for the pulse to pass u, or a time $\Delta u/\epsilon R$. The relevant values of λ will be less than the reciprocal of this time, $\lambda<\epsilon R/\Delta u$. We shall see later that $\Delta u\approx\epsilon^{1/2}$, so $\lambda/R\approx\epsilon^{1/2}$. Accordingly, we expand the integrand in (5.46) in powers of λ/R, and obtain, to second order,

$$S_\lambda=\lambda\int_{u_0}^{u}du'/R'-(2/3)\lambda^2\int_{u_0}^{u}du'/R'^2+const.$$

If we write

$$T=(1/\epsilon)\int_{u_0}^{u}du'/R',$$

which ix the same as (5.37), and

$$(\Delta T)^2=(4/3\epsilon)\int_{u_0}^{u}du'/R'^2, \qquad (5.47)$$

we have

$$w=(C/2\pi i)\int e^{\lambda(t-T)+(1/2)\lambda^2(\Delta T)^2}d\lambda$$

$$=(C/2\pi i)e^{-(1/2)(t-T)^2/(\Delta T)^2}\int e^{(1/2)(\Delta T)^2[\lambda+(T-T)/(\Delta T)^2]^2}d\lambda.$$

Let $\lambda+(t-T)/(\Delta T)^2=ix$, and pick the contour of the λ integration to pass through x=0. Then

$$w=(C/2\pi)e^{-(1/2)(t-T)^2/(\Delta T)^2}\int_{-\infty}^{\infty}e^{-(1/2)(\Delta T)^2x^2}dx$$

$$=\frac{C}{\sqrt{(2\pi(\Delta T)^2)}}e^{-(1/2)(t-T)^2/(\Delta T)^2}. \qquad (5.48)$$

The constant of integration can be determined from the conservation law, $\int_0^\infty q\,dt=1$. Since $\int_0^\infty w\,dt\approx\int_{-\infty}^\infty w\,dt=C$, it follows from (5.42) that the conservation law is satisfied if $C=1/\epsilon$.

The distribution (5.48) can also be expressed as an explicit function of u, rather than t. Let u_t be the average u reached after time t, defined by

$$T(u_t)=t.$$

To second order in $(u-u_t)$,

$$\frac{(t-T)^2}{(\Delta T)^2}=[\frac{1}{(\Delta T)^2}(\frac{dT}{du})^2]_{u=u_t}(u-u_t)^2=\frac{(u-u_t)^2}{\epsilon^2 R^2(\Delta T)^2}\Big|_{u=u_t}.$$

If we define

$$(\Delta u_t)^2=\epsilon^2 R^2(\Delta T)^2\big|_{u=u_t},$$

$$w=\frac{R}{\sqrt{(2\pi(\Delta u_t^2))}}e^{-(1/2)(u-u_t^2)/(\Delta u_t)^2}$$

and

$$\rho(u,t)=n(u,t)=\frac{1}{\sqrt{(2\pi(\Delta u_t)^2)}}e^{-(1/2)(u-u_t)^2/(\Delta u_t)^2}. \qquad (5.49)$$

In deriving (5.49) we have not distinguished between R and $R_{u=u_t}$, that is, we we have supposed the variation of R over the width of the distribution is small. It can readily be checked that this is a necessary condition for neglecting the term on the right in (5.45). Since (5.49) is the asymptotic solution for times long enough for a number of collisions to have taken place, $n_0\approx0$ and $\rho\approx n$, as we have indicated. Note that the conservation law $\int\rho\,du=1$ is satisfied.

If R were a constant independent of energy, we should have

$$(\Delta T)^2=\frac{4(u-u_0)}{3\epsilon R^2},$$

and
$$(\Delta u_t)^2 = (4/3)\epsilon(u_t - u_0) = (4/3)\epsilon^2 N_c = N_c \langle x^2 \rangle,$$

N_c being the mean number of collisions to time t, given by
(5.38), and $\langle x^2 \rangle$ being the mean square spread in u per
collision, given by (5.40). This is just the result we
should expect for statistically independent collisions.
However, if σ is constant we have
$$(\Delta T)^2 = \frac{4}{3\epsilon} \cdot 2 \int_v^{v_0} \frac{1}{(N\sigma)^2} \frac{dv}{v^3} = \frac{4}{3\epsilon}(\frac{1}{R^2} - \frac{1}{R_0^2}),$$

and
$$(\Delta u_t)^2 = (4/3)\epsilon(1 - R^2/R_0^2).$$

In this case $(\Delta u_t)^2$ remains finite in the limit $E_0 \to \infty$. Even
though $N_c \to \infty$, the effective number of collisions approaches
$1/\epsilon$.

In general, the collisions are not statistically
independent. If R increases with energy there is an obvious
bunching effect: neutrons whose energy is higher than the
mean make collisions at a higher rate, lose energy more
rapidly, and tend to approach the mean; neutrons of too low
energy make collisions less rapidly and are overtaken by
the mean.

The solution, $\rho(\mu,t)$ (or $W(u,t)=R\rho(u,t)$), for a source
distributed in time and energy, $S(u_0,t)$, can be obtained by
superposition of the solutions just obtained. If these are
denoted $n(u,u_0,t)$ (or $w(u,u_0,t)=Rn(u,u_0,t)$), we have
$$W(u,t) = \int_{-\infty}^{u} du_0 \int_{-\infty}^{t} dt' w(u,u_0,t-t') S(u_0,t').$$

According to (5.41), w satisfies the equation
$$(1/R)\partial/\partial t + L)w = 0,$$

where L is the linear operator
$$L = \epsilon\partial/\partial u[1 - (2/3)\epsilon\partial/\partial u].$$

We then have
$$[(1/R)\partial/t + L]W(u,t) = \int_{-\infty}^{u} du_0 (1/R)w(u,u_0,0)S(u_0,t) = S(u,t),$$

since, as we see from (5.49), $\lim_{t \to 0} w(u,u_0,t)=R\delta(u-u_0)$.
(The behavior at the limits of integration is clearer if
one takes the upper limit of the t integration to be $t-\delta$,
and lets $\delta \to 0$ at the end.) Thus ρ satisfies (5.08)
(rewritten on the logarithmic scale),

$$\partial\rho/\partial t=-\partial q/\partial u+S, \tag{5.50}$$

with

$$q=\epsilon[W-(2/3)\epsilon\partial W/\partial u]. \tag{5.51}$$

The effect of a capture cross section could be taken into
account by adding a term $-Nv\sigma_{cap}\rho$ on the right side of
(5.50).

SEC.5III SPATIAL DIFFUSION OF NEUTRONS

The previous sections of this chapter have been concerned with the diffusion of neutrons in energy. Their spatial diffusion is also an important problem; for example, if a thermal neutron is at the origin at time 0, what is its probability of being at $\underline{r}$ after time t?

Let us first consider the case of elastic scattering by very heavy targets, so the scattering is spherically symmetric and there is no energy loss. The problem of the distribution after N_c collisions is then a random—walk problem. If $l=1/N\sigma$ is the mean free path, the probability of making a first collision in dr after traversing a distance r is $e^{-r/1}dr/1$. The mean—square distance between collisions is

$$\langle r^2 \rangle = \int_0^{\infty} e^{-r/1} r^2 dr = 21^2,$$

and the mean-squared displacements in the rectangular coordinates are

$$\langle x^2 \rangle = \langle y^2 \rangle = \langle z^2 \rangle = (1/3)\langle r^2 \rangle = (2/3)1^2.$$

For a neutron initially at the origin, we would expect that after a large number of collisions we would find a Gaussian distribution in x, with a mean-square width $(\Delta x)^2 = N_c \langle x^2 \rangle$,

$$\rho(x)dx = 1/\sqrt{(2\pi(\Delta x)^2)} e^{-x^2/2(\Delta x)^2} dx.$$

Multiplying together the distributions in x, y and z, we find

$$\rho(\underline{r})d\underline{r} = (1/2\pi(\Delta x)^2)^{3/2} e^{-r^2/2(\Delta x)^2} d\underline{r}.$$

The time required to make N_c collisions is

$$t = (1/v)N_c,$$

so, in terms of the time,

$$(\Delta x)^2 = (2/3)1vt,$$

and

$$\rho(\underline{r})d\underline{r} = (4\pi 1vt/3)^{-3/2} e^{-3r^2/41vt} d\underline{r}. \qquad (5.52)$$

It is easily checked that this distribution satisfies the diffusion equation

$$\partial\rho/\partial t=(lv/3)\Delta\rho, \qquad (5.53)$$

which can also be written as a conservation equation,

$$\partial\rho/\partial t=-\text{div}\underline{j},$$

with the current, $\underline{j}$, given by

$$\underline{j}=-(lv/3)\text{grad}\rho. \qquad (5.54)$$

A generalization of (5.53), in which we include a capture cross section, σ_c, as well as the elastic cross section σ, and also a neutron source of S neutrons/cc/sec, is

$$\partial\rho/\partial t=(lv/3)\Delta\rho-Nv\sigma_c+S. \qquad (5.55)$$

The argument leading to (5.52) and (5.53) depended on the number of collisions, N_c, being large, or, equivalently, on the distribution changing little in a mean free path. This is evidently not compatible with a large capture cross section, so (5.55) can be correct only in the limit $\sigma_c/\sigma\ll1$.

Some insight into the limitations of the diffusion equation, as well as suggestions for improving it, can be obtained by considering the case of the static solution, with no source, and ρ varying only in one direction.

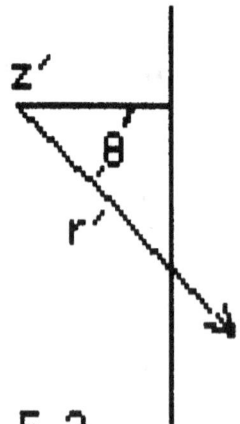

Fig.5.2

Equation (5.55) becomes

$$(1/3N\sigma)d^2\rho/dz^2=N\sigma_c\rho,$$

and the solution is

$$\rho=Ae^{s_0z}+Be^{-s_0z},$$

with

$$s_0=N\sqrt{(3\sigma\sigma_c)}. \qquad (5.56)$$

Now let us remove the restriction $\sigma_c/\sigma\ll1$. Let σ_t be the total cross section, $\sigma_t=\sigma+\sigma_c$, and $\lambda=\sigma/\sigma_t$, $1-\lambda=\sigma_c/\sigma_t$. Consider the flux of neutrons crossing the z=0 plane from the left, in the angular range $d\mu$ ($\mu=\cos\theta$). In

Fig 5.2, z' is the point at which the neutron makes its last collision. The flux is

$$Fd\mu=(Nv\sigma/4\pi)\int_{-\infty}^{0}\rho(z')e^{-N\sigma_t r'}\,dz'\cdot 2\pi d\mu.$$

Let us measure distances in units $1/N\sigma_t$,

$$z=N\sigma_t z', \quad r=N\sigma_t r'.$$

Then

$$F=(v\lambda/2)\int_{-\infty}^{0}e^{-r}\rho(z)dz=(v\lambda/2)\int_{-\infty}^{0}e^{z/\mu}\rho(z)dz.$$

If

$$\rho(z)=\rho_s e^{sz} \qquad (5.57)$$

we find

$$F_s=(1/2)v\lambda\rho_s\mu/(1+s\mu). \qquad (5.58)$$

Similarly, the flux (defined to be negative) coming from the right is

$$F=-(v\lambda/2)\int_{0}^{\infty}e^{z/\mu}\rho(z)dz. \qquad (5.59)$$

Note that θ is still measured from the positive z axis, so in this case μ is negative. The same result, (5.58), is obtained (provided $s<-1/\mu$ so the integral converges), so (5.58) holds over the whole angular range.

If $n(z,\mu)d\mu$ is the density of neutrons moving in direction θ, $F=v_z n=v\mu n$, so

$$n_s=(1/2)\lambda\rho_s(1+s\mu). \qquad (5.60)$$

The current crossing the $z=0$ plane to the right is

$$j_+=\int_{0}^{1}F_s d\mu=(1/2s^2)\lambda v\rho_s[s-\ln(1+s)],$$

that to the left is

$$j_-=\int_{-1}^{0}F_s d\mu=(1/2s^2)\lambda v\rho_s[s+\ln(1-s)], \qquad (5.61)$$

and the net current is

$$j=\int_{-1}^{1}F_s d\mu=j_++j_-=(1/2s^2)\lambda v\rho_s[2s-\ln\{(1+s)/(1-s)\}].$$

Furthermore,

$$\rho=\int_{-1}^{1}nd\mu; \qquad (5.62)$$

using (5.58) for n gives a relationship between s and λ,

$$\lambda = 2s/\ln\frac{(1+s)}{(1-s)}. \qquad (5.63)$$

The diffusion equation holds in the region where ρ changes little in a mean free path, which means $|s| \ll 1$. On expanding (5.61) as a power series in s we find, to lowest order in s,

$$j_+ = (1/2)\lambda v \rho_s ((1/2) - (1/3)s),$$
$$j_- = (1/2)\lambda v \rho_s (-(1/2) - (1/3)s), \qquad (5.64)$$
$$j = -(\lambda v/3)s \rho_s.$$

For small s (5.63) can be expanded in the form
$$(1/2s)\ln(1+s)/(1-s) = 1 + (1/3)s^2 + (1/5)s^4 + \ldots = 1/\lambda.$$
To lowest order in s^2 there are two roots, $\pm s_0$, with

$$s_0 = \sqrt{[3(1-\lambda)/\lambda]} = \sqrt{(3\sigma_c/\sigma)}. \qquad (5.65)$$

In our present units, (5.56) reads $s_0 = \sqrt{[3\lambda(1-\lambda)]}$. Since λ differs from unity only to order s^2 the difference between (5.56) and (5.65) is not really significant, although it may be remarked that (5.56) gives a better approximation than (5.65) to the roots of (5.63). The last of equations (5.64) can be written, in view of (5.57) as

$$j = -(\lambda v/3)d\rho(0)/dz, \qquad (5.66)$$

in agreement with (5.54) to lowest order. In the opposite limit, $\sigma \to 0$, $\lambda \to 0$, (5.63) gives s=1, which is the correct answer.

Let us now consider what happens at the surface of a scattering medium: suppose the scatterer occupies only the space to the left of z=0 (Fig 5.3a). The neutron density in the scatterer will be the same as in the infinite medium, provided there is an incoming flux of neutrons, moving to the left, which is the same as the characteristic distribution, (5.58). In this case the neutron flux and distribution in the space z>0 will be the same as F_s and n_s at z=0 in the infinite medium. Such a situation could be

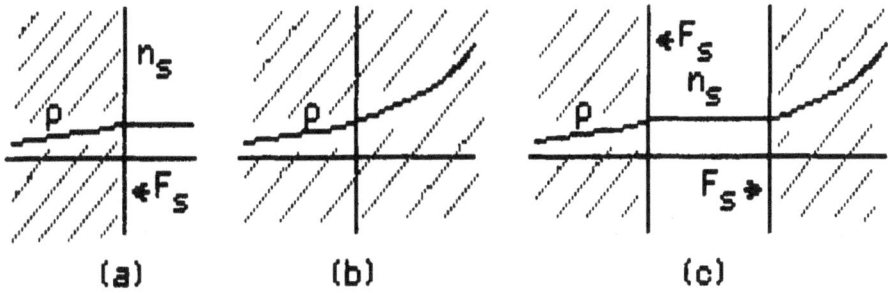

Fig.5.3

obtained physically by splitting the infinite medium
(Fig 5.3b) at the z=0 plane, and creating a gap (Fig 5.3c).

   The above considerations lead to a particular solution
of the albedo problem. The albedo is the reflection
coefficient for a neutron striking the surface. The albedo
depends, in general, on the angle of incidence of the
neutron, If the angular distribution of incidence is that
of the characteristic distribution the albedo is just

$$A_s = \frac{j_+}{j_-} = \frac{\ln(1+s)-s}{\ln(1-s)+s} .$$

For small s, as we see from (5.64)

$$A_s = \frac{1-(2/3)s_0}{1+(2/3)s_0} . \tag{5.67}$$

   To first order in s, (5.67) is also the albedo for
n=const on the right, i.e. for an incoming flux $F \approx \mu$, since
the characteristic distribution differs from n=const only
by a term proportional to s_0, and of this only a fraction
again proportional to s_0 will be absorbed in the medium, as
is evident from (5.67). The same result can be obtained
from the solution, (5.56), of the diffusion equation. If
to the right of the surface $n(\mu)=n_0$, we have

$$\rho_0 = \int_{-1}^{1} n_0 d\mu = 2n_0, \qquad j_- = v \int_{-1}^{0} n_0 \mu d\mu = -(1/2)vn_0.$$

Within the medium the solution vanishing at $z=-\infty$ is

$$\rho = \rho_0 e^{s_0 z} = -(4j_-/v)e^{s_0 z},$$

and the net ingoing current at the surface is, by (5.66)

$$-j = (4s_0/3)j_-.$$

The fraction of incoming neutrons absorbed by the medium is $-j/j_- = 4s_0/3$, so the albedo is

$$A = 1 - 4s_0/3,$$

in agreement with (5.67)

A more general treatment of the problem is obtained by considering the Boltzman equation for the phase space distribution, $n(\underline{r},\underline{v},t)$. For elastic scattering (i.e. no energy loss) the magnitude of v is fixed but its direction varies over the surface of a sphere. (The distribution $n(z,\mu)$ we considered for one-dimensional flow is a special case.) The angular distribution of the scattering may be arbitrary,

$$d\sigma = \sigma f(\cos\theta)d\Omega, \qquad \int f(\cos\theta)d\Omega = 1.$$

The distribution satisfies the equation

$$\partial n(\underline{r},\underline{v},t)/\partial t = -\underline{v}\cdot\text{grad } n(\underline{r},\underline{v},t) - Nv\sigma_t n(\underline{r},\underline{v},t)$$
$$+ \int Nv\sigma f(\cos\psi) n(\underline{r},\underline{v}',t)d\Omega_{\underline{v}'} + S(\underline{r},\underline{v},t) \qquad (5.68)$$

The first term on the right can also be written $-\text{div}(\underline{v}n)$; it is minus the divergence of the current, and gives the change of n at $\underline{r}$ due to the flux of neutrons across the surface of an element of volume surrounding $\underline{r}$. The second term gives the rate at which neutrons are removed from $\underline{v}$ and the third term gives the rate of scattering to $\underline{v}$ from $\underline{v}'$, ψ being the angle between $\underline{v}$ and $\underline{v}'$. The term $S(\underline{r},\underline{v},t)$ represents any source which may be present.

For the case we have been considering, unidirectional flow and spherically symmetric scattering, we have $f(\cos\theta) = 1/4\pi$ and (5.68) becomes

$$\partial n(z,\mu,t)/\partial t = -v\mu \partial n(z,\mu,t)/\partial z - N v \sigma_t n(z,\mu,t)$$

$$+(1/2)Nv\sigma \int_{-1}^{1} n(z,\mu',t)d\mu' + S(z,\mu,t)$$

If we measure lengths in units $1/N\sigma_t$ and time in units $1/Nv\sigma_t$ (which amounts to putting $N\sigma_t = v = 1$ and $N\sigma = \lambda$ in the above equation we obtain (remembering (5.62))

$$\partial n(z,\mu,t)/\partial t = -\mu \partial n(z,\mu,t)/\partial z - n(z,\mu,t) + (1/2)\lambda \rho(z,t) + S(z,\mu,t).$$

$$(5.69)$$

Let us consider the case in which the scatterer fills the half space to the right of $z=0$. Then, in addition to the volume source S, we must consider the source (or sink) due to the flux of neutrons crossing the $z=0$ plane. Since we will express the z-dependence of the solutions in terms of Laplace integrals, it is mathematically convenient to seek solutions for which n and ρ are zero for $z<0$. There will then be a step function discontinuity of $n(z,\mu,t)$ at $z=0$, so $\partial n/\partial z$ will have a δ function discontinuity, $n(0,\mu,t)\delta(z)$, and (5.69) becomes

$$\partial n(z,\mu,t)/\partial t = -\mu \partial n(z,\mu,t)/\partial z - n(z,\mu,t) + (1/2)\lambda \rho(z,t)$$

$$+S(z,\mu,t) + F(\mu,t)\delta(z), \qquad (5.70)$$

where

$$F(\mu,t) = \mu n(0,\mu,t)$$

is the flux across the $z=0$ plane.

If we again look at the static case with $S(z,\mu)=0$, we obtain for the Laplace transform of (5.70)

$$(1+s\mu)n_s = (1/2)\lambda \rho_s + F(\mu),$$

so

$$n_s = \frac{1}{2}\frac{\lambda \rho_s}{1+s\mu} + \frac{F(\mu)}{1+s\mu}, \qquad (5.71)$$

which differs from (5.60) by the surface flux term. For $\mu>0$, $F(\mu)$ is a known function, being the flux directed at the surface of the scatterer. For $\mu<0$, $F(\mu)$ is the flux emitted from the surface of the scatterer, which is given by (5.59)(with $v=1$, in light of our present choice of units). By the general rule for Laplace transforms, (5.11), we thus have

$$F(\mu)=-(1/2)\lambda\rho_{-1/\mu}, \qquad \mu<0. \qquad (5.72)$$

Note that for $\mu<0$, (5.71) is

$$n_s=(1/2)\lambda(\rho_s-\rho_{-1/\mu})/(1+s\mu),$$

so the apparent pole in the right hand plane at $s=-1/\mu$ is removed by a zero of the numerator.

If (5.71) is integrated over μ we obtain, since

$$\rho_s=\int_{-1}^{1} n_s d\mu,$$

$$\rho_s=\frac{\lambda}{2s}\rho_s \ln\frac{1+s}{1-s} +\int_{-1}^{1} F(\mu)d\mu/(1+s\mu),$$

or, solving for ρ_s,

$$\rho_s=\frac{\displaystyle\int_{-1}^{1} F(\mu)d\mu/(1+s\mu)}{1-\dfrac{\lambda}{2s}\ln\dfrac{1+s}{1-s}}. \qquad (5.73)$$

This, in view of (5.72), is an integral equation for ρ_s.

Equation (5.73) has poles at the points where (5.63) is satisfied. Thus, in the diffusion equation limit, when $1-\lambda\ll1$, there are poles for s near the origin, at $\pm s_0$. The density, $\rho(z)$, is given by

$$\rho(z)=(1/2\pi i)\int_C \rho_s e^{sz}ds, \qquad (5.74)$$

the contour passing to the right of the poles. For large z the dominant contributions will come from the residues at $\pm s_0$. These will give a result of the form, (5.56), of a solution of the diffusion equation. In physical terms, this means that (5.56) holds within the interior of the medium, but within distances of the order of a mean free path of the surface the remainder of the integral (5.74) may make significant changes.

If we expand (5.73) in powers of s we find to lowest order

$$\rho_s = \frac{\int_{-1}^{1} F(\mu)(1-s\mu)\,d\mu}{1-\lambda(1+s^2/3)},$$

which gives for (5.74)

$$\rho(z) \approx -\frac{3}{2\pi i}\int_C \frac{(j-sj_1)}{s^2-s_0^2}e^{sz}\,ds,$$

where j is the net current across the surface,

$$j = \int_{-1}^{1} F(\mu)\,d\mu,$$

and j_1 is the first moment,

$$j_1 = \int_{-1}^{1} \mu F(\mu)\,d\mu.$$

Evaluating this in terms of the residues at the poles gives

$$\rho(z) \approx (3/2s_0)[(j+s_0 j_1)e^{-s_0 z} - (j-s_0 j_1)e^{s_0 z}]. \qquad (5.75)$$

Note that at z=0 the diffusion theory relationship, $j=-(1/3)d\rho/dz$, is satisfied.

As an example, let us try to calculate the albedo for the case when, on the left, neutrons are coming in at a definite angle, θ_0. We can then write

$$n(\mu)=\delta(\mu-\mu_0), \quad F(\mu)=\mu\delta(\mu-\mu_0), \quad \mu>0.$$

Within the medium the density must be falling off for large z, so we have only the $s=-s_0$ term. We suppose that, for $\mu<0$, $F(\mu)$ is approximately of the form of the characteristic distribution, (5.58), and write

$$F(\mu)=2\mu_0 A\mu/(1-s_0\mu) \approx 2\mu\mu_0 A(1+s_0\mu), \quad \mu<0,$$

to first order in s_0. The constant A is determined by the condition that the exponentially increasing term in in (5.75) vanishes.

The incident and reflected currents are

$$j_+ = \int_0^1 F d\mu = \mu_0, \quad j_- = \int_{-1}^0 F d\mu = -\mu_0 A(1-2s_0/3),$$

and the net current is

$$j = j_+ + j_- = \mu_0[1-A(1-2s_0/3)].$$

To zeroth order in s_0,

$$j_1 = \mu_0^2 + (2/3)\mu_0 A = \mu_0(\mu_0 + 2A/3).$$

The condition $j-s_0 j_1=0$ gives
$$A=1-s_0\mu_0;$$
with this value
$$j_-=-\mu_0[1-s_0(\mu_0+2/3)], \quad j=s_0\mu_0(\mu_0+2/3),$$
and (5.75) gives
$$\rho(z)=3j/s_0 e^{-s_0 z}=3\mu_0(\mu_0+2/3)e^{-s_0 z}. \tag{5.76}$$

The albedo is
$$A(\mu_0)=-j_-/j_+=1-s_0(\mu_0+2/3).$$
Our previous result for the albedo when n on the left is
uniformly distributed in angle can be obtained by
averaging:
$$\langle A \rangle = \frac{-\int_0^1 j_-(\mu_0)d\mu_0}{\int_0^1 j_+(\mu_0)d\mu_0} = \frac{\int_0^1 \mu_0 A(\mu_0)d\mu_0}{\int_0^1 \mu_0 d\mu_0} = 1-4s_0/3.$$

Equation (5.76) gives
$$\rho(0)=3\mu_0(\mu_0+2/3). \tag{5.77}$$

However, ρ_0 can also be calculated from the relation
$$\rho(0)=\int_{-1}^1 n d\mu = \int_{-1}^1 (F/\mu)d\mu.$$
This gives, to zeroth order in s_0,
$$\rho(0)=1+2\mu_0. \tag{5.78}$$

Equation (5.78) gives $\rho(0)$ larger than (5.77) for $\mu_0<1/\sqrt{3}$
$(\theta_0>71°)$ and $\rho(0)$ smaller for $\mu_0>1/\sqrt{3}$. It must be
remembered that (5.76) gives the density at interior points
of the scatterer; thus (5.77) gives the extrapolation of
the asymptotic solution to the surface. We may suppose that
the true situation is something like that illustrated in
Fig 5.4 for the case $\mu_0=1$, where (5.77) gives $\rho(0)=5$ and
(5.78) gives $\rho(0)=3$.

The outgoing flux is
$$F(\mu,\mu_0)=2\mu_0\mu[1-s_0(\mu_0-\mu)], \quad \mu<0, \quad \mu_0>0.$$
This satisfies the reciprocity (or detailed balancing)
relation,

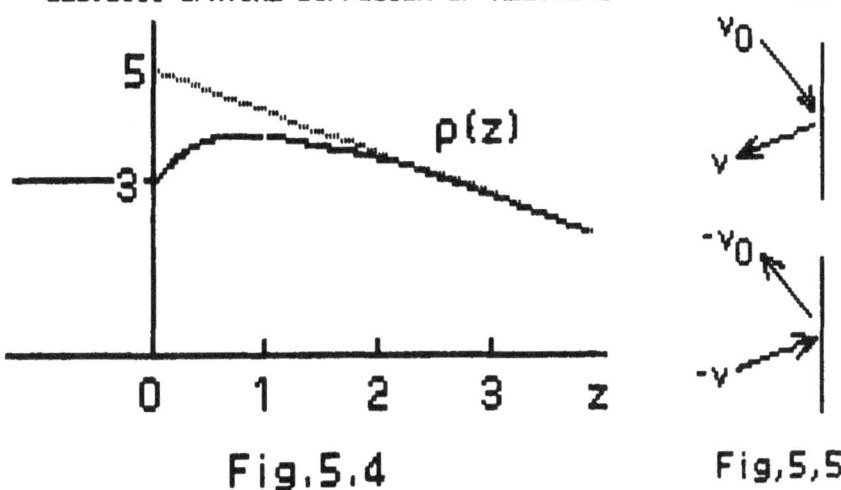

Fig.5.4   Fig,5,5

$$F(-\mu_0,-\mu)=F(\mu,\mu_0),$$

illustrated in Fig 5.5, i.e. the reflection coefficient is
unchanged if the velocities are reversed, so that the roles
of incoming and outgoing particles are reversed. The
requirement of the reciprocity relation could be used as
an alternative method of solving the albedo problem.

    Another important problem is that of a surface with no
incoming flux. Suppose that in the scatterer $(z>0)$ the
density of neutrons is

$$\rho(z)\approx\rho_{s_0}e^{s_0 z}+\rho_{-s_0}e^{-s_0 z}.$$

For $z<<1/s_0$

$$\rho(z)\approx(\rho_{s_0}+\rho_{-s_0})+s_0(\rho_{s_0}-\rho_{-s_0})z=\rho_0-3jz, \qquad (5.79)$$

where ρ_0 is the extrapolation of the asymptotic form to $z=0$
and $j=-(1/3)d\rho/dz$ is the current in the scatterer. We
readily find

$$\rho_{s_0}=(1/2)(\rho_0-3j/s_0), \quad \rho_{-s_0}=(1/2)(\rho_0+3j/s_0).$$

If we supposed (5.79) held right to the surface, the
outgoing current at the surface would be given by (5.64),
which now reads (remember $v=1$ in our present units)

$$j_-=(1/2)\lambda[(1/2)(\rho_0-3j/s_0)(-(1/2)-s_0/3)$$
$$+(1/2)(\rho_0+3j/s_0)(-(1/2)+s_0/3)].$$

To lowest order in s_0 this gives
$$j_- = (1/4)(-\rho_0 + 2j).$$
Conservation requires $j_- = j$, which gives the condition
$$\rho_0 = -2j. \tag{5.80}$$

With this condition, (5.79) can be written
$$\rho(z) \approx -3j(z + z_0), \tag{5.81}$$

with $z_0 = 2/3$. Thus we obtain Fermi's algorithm that the asymptotic density extrapolates to zero 2/3 of a mean free path outside the surface of the scattering medium.

The angular distribution is similarly given by (5.60); the result is
$$n(0,\mu) = (1/2)(\rho_0 + 3j\mu), \quad \mu < 0.$$
Inserting (5.80) gives
$$n(0,\mu) = -j(1 - 3\mu/2), \quad \mu < 0. \tag{5.82}$$
The density at the surface is
$$\rho(0) = \int_{-1}^{0} n(0,\mu)\,d\mu = -3j \cdot 7/12 = -3j \cdot 0.5833, \tag{5.83}$$

which is less than the extrapolated value, $\rho_0 = -3jz_0 = -3j \cdot 0.6667$. The density thus looks like Fig 5.7, which is drawn for $d\rho/dz = -3j = 1$. Numerical values are given in Table 5.1. According to (5.82) and (5.80) $n(0,0) = (1/2)\rho_0$. In fact, the relationship
$$n(0,0) = (1/2)\lambda\rho(0) \tag{5.84}$$
is exact, as can be seen to follow from (5.59). The point is that, as $-\mu \to 0$, the scattering must take place at the surface. The difference between $n(0,0) = 0.3333$, given

TABLE 5.1 $(d\rho/dz = -3j = 1)$				
	From (5.82)	(a) From (5.75) (b) From (5.84)	Weiner–Hopf Solution	From (5.86)
z_0	0.6667	0.7083 (a)	0.7104	0.7113
$\rho(0)$	0.5833		0.5774	0.5774
$n(0,-1)$	0.8333		0.8394	0.8453
$n(0,0)$	0.3333	0.2917 (b)	0.2887	0.3094

by (5.82) for $-3j=1$, and $n(0,0)=.2917$, given by (5.84) and (5.83) indicates some inaccuracy in (5.82) at tangential angles.

We have so far made no use of (5.75), which gives

$$\rho_0 = 3j_1, \text{ or } z_0 = -j_1/j. \tag{5.85}$$

If we use our approximate $F(\mu)$, from (5.82), we obtain

$$z_0 = \int_{-1}^{0} (\mu^2 - 3\mu^3/2)\,d\mu = 17/24 = 0.7083$$

as a better estimate of z_0 than $z_0 = 2/3$.

An exact solution of the integral equation (5.73) has been given for $\lambda=1$ (the Wiener-Hopf solution)[3]. The exact results are, as shown in Table 5.1, $z_0 = 0.7014$, $\rho(0) = 1/\sqrt{3} = 0.5774$, $n(0,-1) = 0.8394$, $n(0,0) = 0.2887$. The comparison between the exact $n(0,\mu)$ and (5.82) is shown in Fig 5.6, as is also another linear approximation for $n(0,\mu)$,

$$n(0,\mu) = 0.3094(1 - \sqrt{3}\mu). \tag{5.86}$$

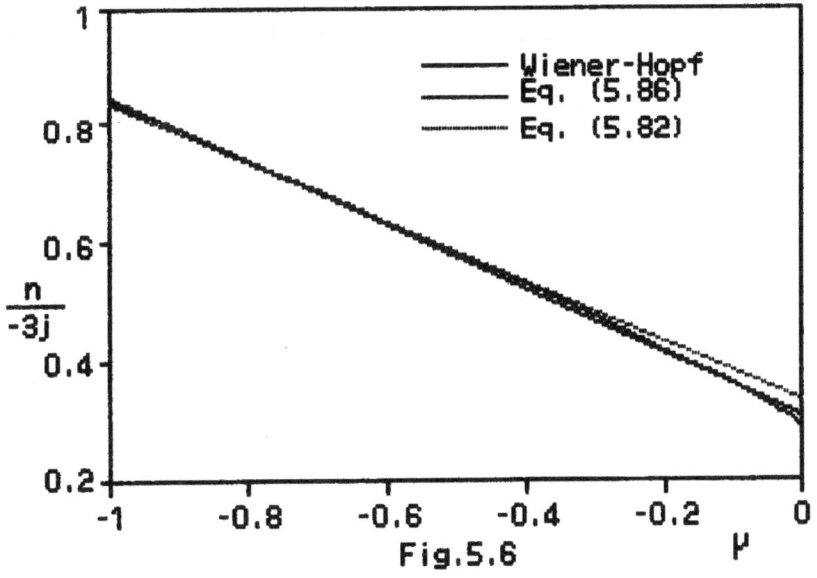

Fig.5.6

---

[3] See G. Placzek, Phys. Rev. __72__, 556 (1947).

This approximation, which is due to Fermi[3], can be obtained by using the conservation condition, $j_- = j$, and the Wiener-Hopf result, $\rho(0) = 1/\sqrt{3}$.   If we write

$$n(0,\mu) = C(1-a\mu),$$

we have (for $-3j=1$),

$$j_- = C\int_{-1}^{0}(\mu - a\mu^2)\,d\mu = -(C/2)(1+2a/3) = -1/3,$$

$$\rho(0) = C\int_{-1}^{0}(1-a\mu)\,d\mu = C(1+a/2) = 1/\sqrt{3}.$$

Solving for C and a gives

$$C = 1/[\sqrt{3}(1+\sqrt{3}/2)] = 0.3094, \quad a = \sqrt{3}.$$

Equation (5.85) then gives

$$z_0 = 3C\int_{-1}^{0}(\mu^2 - \sqrt{3}\mu^3)\,d\mu = 3C(1/3 + \sqrt{3}/4) = 0.7113.$$

The failure of the linear approximations for $n(0,\mu)$ near $-\mu=0$ is due to the fact that the exact solution is singular at $-\mu=0$; for very small values of $-\mu$

$$n(0,\mu) = (1/2\sqrt{3})[1+(1/2)\mu\ln(-\mu)].$$

The exact solution for $\rho(z)$ shows a similar behavior as $z\to0$,

$$\rho(z) = (1/\sqrt{3})[1-(1/2)z\ln z].$$

Fig.5.7 shows the exact solution for $\rho(z)$[4]. The straight line is the solution of the diffusion equation, $\rho(z)=z+z_0$, with $z_0=0.7104$.

Another interesting question concerns the modification of the diffusion equation if the scattering is not spherically symmetric. Returning to (5.68), we can, at point r, choose the z axis in the direction of grad n, and write $n=n(r,\mu,t)$ with, as usual, $\mu=\cos\theta$ and θ the angle between v and the z axis. Remembering that $\sigma_t = \sigma + \sigma_c$, we see that the elastic cross section, σ, occurs in (5.68) in the two terms

$$-Nv\sigma n(r,\mu,t) + Nv\sigma\int f(\cos\psi)n(r,\mu',t)\,d\Omega'. \qquad (5.87)$$

---

[4] C. Mark, Phys. Rev. 72, 558 (1947).

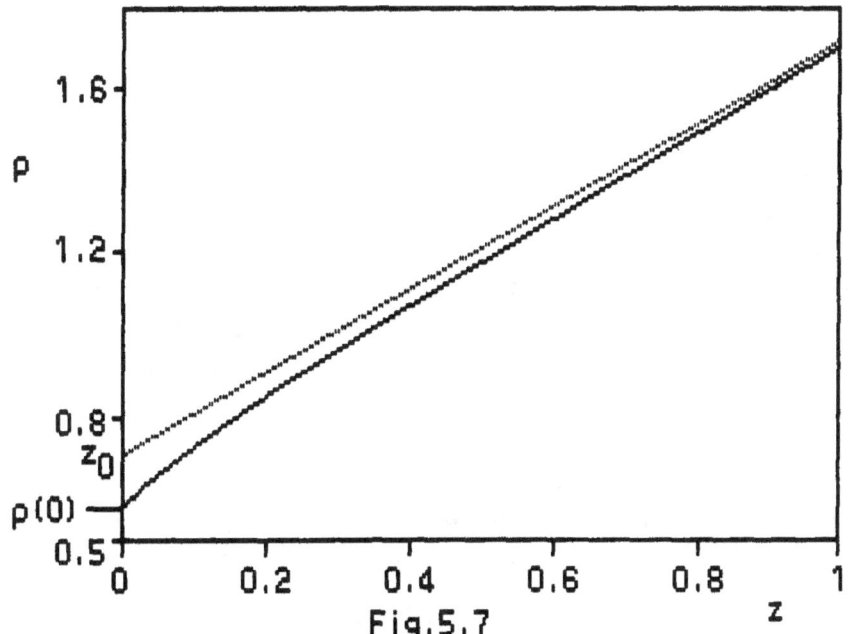

Fig.5.7

We now make the approximation appropriate to the diffusion equation, that n is a linear function of μ,

$$n = n_0 + n_1 \mu. \tag{5.88}$$

In the spherical triangle with sides θ, θ' and ψ,

$$\mu' = \cos\theta' = \cos\psi\cos\theta + \sin\psi\sin\theta\cos\xi,$$

with ξ the angle between ψ and θ. This gives, when (5.88) is inserted in (5.87),

$$-Nv\sigma(n_0 + n_1\mu) + Nv\sigma(n_0 + n_1\langle\cos\psi\rangle\mu) = -Nv\sigma(1 - \langle\cos\psi\rangle)n_1.$$

Thus σ appears in (5.68) only in the combination $\sigma(1 - \langle\cos\psi\rangle)$. This means that in the diffusion equation the elastic scattering cross section, σ, should be replaced by the so-called transport cross section,

$$\sigma_{tr} = \sigma(1 - \langle\cos\psi\rangle).$$

If the scattering is predominantly forward, so $\langle\cos\psi\rangle$ is positive, the effective mean free path is correspondingly increased.

The calculation we have just done amounts to considering the first two terms of an expansion of n as a series of spherical harmonics and can be improved by considering higher terms, a method developed by R. Marshak. Another way that can sometimes be employed to improve solutions of the diffusion equation is to use (5.63) instead of (5.65) to determine s_0. When the scattering medium contains fissionable material, so neutrons may be produced rather than absorbed, $1-\lambda$ changes sign, so $\lambda>1$. In this case s becomes imaginary, $s=ik$, and (5.63) becomes

$$\lambda=k/\tan^{-1}k.$$

Finally, we may combine the results of this section and Sec 5II to obtain the equation for $\rho(\underline{r},u,t)$,

$$\partial\rho/\partial t=-\partial q/\partial u-\text{div }j-Nv\sigma_c+S, \qquad (5.89)$$

with q given by (5.51) and $\underline{j}$ by $\underline{j}=-(1/3)l_{tr}v\text{grad }\rho$, where $l_{tr}=1/N\sigma_{tr}$.

In the case of a source of neutrons of energy E_0, which is constant in time, and with $\sigma_c=0$, (5.89) can be written, for $E<E_0$,

$$\partial q/\partial u=(1/3)ll_{tr}\Delta W. \qquad (5.90)$$

Fermi's "age" equation is obtained by neglecting the dispersion of neutron energies in a collision, i.e. neglecting the ϵ^2 proportional term in (5.51) and taking $q=\epsilon W$, so (5.90) becomes

$$\partial q/\partial u=(ll_{tr}/3\epsilon)\Delta q.$$

Fermi defines the "age" as

$$\tau=\int_{u_0}^{u}(l'l_{tr}'/3\epsilon)du',$$

and the equation becomes

$$\partial q/\partial\tau=\Delta q.$$

For a point source of one neutron per second at the origin the solution is (see (5.53) and (5.52))

$$q=(4\pi\tau)^{-3/2}e^{-r^2/4\tau}.$$

CHAPTER 6

NUCLEON MAGNETIC MOMENTS
AND QUADRUPOLE MOMENT
OF THE DEUTERON

SEC.6.1 MAGNETIC MOMENT OF THE PROTON

Estermann and Stern[1] made the first measurement of the proton magnetic moment by deflecting hydrogen molecules in an inhomogeneous magnetic field, and found the anomalous value $\mu_P = 2.5\ \mu_0 \pm 10\%$ ($\mu_0 = e/2M_P$). Much higher precision can be obtained using Rabi's molecular beam magnetic resonance method, illustrated in Fig 6.1. The molecules in the beam can exist in a number of states, i, with different magnetic moments. In the A magnet, which has an inhomogeneous magnetic field, the molecules are given a deflection proportional to their magnetic moments. They then traverse the C magnet, which has a constant magnetic field. The inhomogeneous field in the B magnet deflects them in the opposite direction and causes them to strike the detector. In the C magnet there is also a small oscillating magnetic field, perpendicular to $\underline{H}$. If the frequency of this field satisfies

$$\pm \omega = E_i(H) - E_j(H), \tag{6.01}$$

where $E_i(H)$ is the energy of the molecule in state i,

---

[1] I. Estermann and O. Stern, Zeits f. Physik 85, 17 (1933)

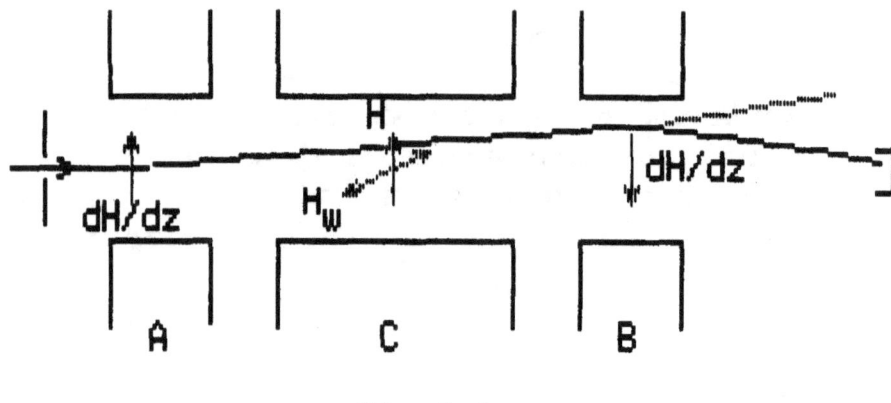

Fig.6.1

transitions will be caused between states i and j, and if
the magnetic moments of states i and j are different the
deflections in magnet B will be changed and the molecules
will no longer strike the detector, as illustrated by the
lighter line in Fig 6.1. Thus if H is varied the detected
intensity will dip at the point where (6.01) is satisfied.

Rabi and his collaborators[2] carried out experiments
for molecules of H_2, HD and D_2.

The simplest case is that of molecules with rotational
quantum number J=0. This excludes H_2; the J≠0 state has the
sum of the two proton spins coupled to S=0 (parahydrogen),
and the magnetic moment is zero. For HD in the J=0 state
the magnetic energy is

$$V = -2\mu_P \underline{S}_P \cdot \underline{H} - \mu_D \underline{S}_D \cdot \underline{H} = -2\mu_P H m_{SP} - \mu_D H m_{SD},$$

m_{SP} and m_{SD} being the magnetic quantum numbers for proton
and deuteron. Under the influence of the oscillating field

---

[2] J.M.B. Kellog, I.I. Rabi and J.R. Zacharias, Phys. Rev.
50, 472 (1936); J.M.B. Kellog, I.I. Rabi, N.F. Ramsey and
J.R. Zacharias, Phys. Rev. 56, 728 (1939).

the m's can change by ± 1, so the resonance condition is
$$\omega = 2\mu_P H \quad \text{or} \quad \omega = \mu_D H.$$

For D_2, since the deuterons obey Bose statistics, the molecule in the $J=0$ state must have a spin function even under deuteron exchange. This is true if the total spin is 2 or 0 (the $S=1$ state is odd). The $S=2$ state has a magnetic energy
$$V = -\mu_D (\underline{S}_1 + \underline{S}_2) \cdot \underline{H} = -\mu_D \underline{S} \cdot \underline{H} = -\mu_D H M_s ,$$
again giving a resonance condition
$$\omega = \mu_D H.$$

By measuring the resonance H for a known ω, Rabi and his collaborators obtained the values
$$\mu_P = 2.785 \pm 0.02 \ \mu_0$$
and
$$\mu_D = 0.855 \pm 0.006 \ \mu_0.$$

Ortho H_2 was measured in the $J=1$ state. The magnetic energy in this case is
$$V = -\mu_P (\underline{\sigma}_1 + \underline{\sigma}_2) \cdot \underline{H} - \mu_R \underline{J} \cdot \underline{H} - \mu_P H' (\underline{\sigma}_1 + \underline{\sigma}_2) \cdot \underline{J}$$
$$- \frac{\mu_P}{r^3} \frac{[3(\underline{\sigma}_1 \cdot \underline{r})(\underline{\sigma}_2 \cdot \underline{r}) - (\underline{\sigma}_1 \cdot \underline{\sigma}_2) r^2]}{r^2}$$
$$= -2\mu_P \underline{S} \cdot \underline{H} - \mu_R \underline{J} \cdot \underline{H} - 2\mu_P H' \underline{S} \cdot \underline{J} - 2 \frac{\mu_P}{r^3} \frac{[(\underline{S} \cdot \underline{r})^2 - \underline{S}^2 r^2]}{r^2}. \qquad (6.02)$$

The first term is the interaction of the proton magnetic moments with the external magnetic field. The second is the orbital interaction with the external field. The third is the spin orbit interaction, $H' \underline{J}$ being the magnetic field seen by the protons as a result of the rotation of the molecule. The last term is the dipole-dipole magnetic interaction, with $\underline{r}$ the separation of the two protons. It can easily be verified that the form in which this interaction is written in the second version of (6.02) equals that in the first: in the second form write $\underline{S} = (1/2)(\underline{\sigma}_1 + \underline{\sigma}_2)$ and multiply out.

The problem is similar to the anomalous Zeeman effect in atoms. There are two limiting cases in which the answer is simple. The strong field case is that in which the external field, H, is large compared to the intrinsic

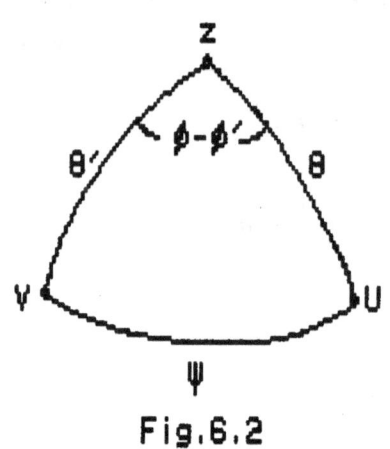

Fig.6.2

magnetic fields. The answer is then given by the diagonal matrix elements of V in an M_S, M_J representation. In the weak field case, H small compared to the intrinsic fields, the spin and orbital angular momentum are coupled to a resultant $\underline{F}=\underline{S}+\underline{J}$, and one requires the diagonal matrix elements of V in an F,M_F representation.

In either case, the treatment of the last term in (6.02) is less familiar than that of the Zeeman terms, and must be discussed. The factor $3(\underline{S}\cdot\underline{r})^2-\underline{S}^2r^2$ is a scalar under rotations, but behaves like a second spherical harmonic if $\underline{r}$ or $\underline{S}$ is rotated separately. To understand precisely what this statement means, consider a case where the meaning is evident, $3(\underline{U}\cdot\underline{V})^2-U^2V^2$, where $\underline{U}$ and $\underline{V}$ are c-number vectors. We can write

$$(1/2)[3(\underline{U}\cdot\underline{V})^2-U^2V^2]=U^2V^2P_2(\cos\psi)$$
$$=U^2V^2\Sigma_\sigma(-1)^\sigma P_2^\sigma(\cos\theta)P_2^{-\sigma}(\cos\theta')e^{i\sigma(\emptyset-\emptyset')}, \qquad (6.03)$$

The angles are shown in Fig 6.2, $(\theta,\emptyset)$ and $(\theta',\emptyset')$ being measured in a fixed coordinate system. Note that in the final form $U^2P_2^\sigma(\cos\theta)e^{i\sigma\emptyset}$ and $V^2 P_2^{-\sigma}(\cos\theta')e^{-i\sigma\emptyset'}$ are polynomials of degree 2 in the components V_i and U_i respectively, polynomials which transform under rotations as do the spherical harmonics, and (6.03) is an algebraic identity in the components of the vectors. Further, we note

that from the nine products $U_i V_j$ we can construct a scalar,
$\underline{U} \cdot \underline{V}$, a vector, $\underline{U} \times \underline{V}$, and the five components of a traceless
symmetric tensor, while from the six products $U_i U_j$ we
obtain the scalar and the symmetric tensor. The five
polynomials $U^2 P_2^\sigma (\cos\theta) e^{i\sigma\emptyset}$ are linearly independent
combinations of the components of the symmetric tensor. The
last term in (6.03) is the scalar obtained by contraction
of the two symmetric tensors formed from $U_i U_j$ and $V_i V_j$.

If we now take $\underline{U} = \underline{r}$, $\underline{V} = \underline{S}$, the components of $\underline{V}$ no longer
commuting, there might be some ambiguity in the meaning of
"$P_2 (\cos\psi)$", but the algebraic identity with the final form
remains true, provided we write the polynomials
$S^2 P_2^{-\sigma} (\cos\theta') e^{-i\sigma\emptyset'}$ in terms of the symmetric tensor, i.e.
$(1/2)(S_x S_z + S_z S_x)$ rather than $S_x S_z$ or $S_z S_x$. The form

$$(1/2)[3(\underline{S} \cdot \underline{r})^2 - \underline{S}^2 r^2] =$$
$$\Sigma_\sigma (-1)^\sigma S^2 P_2^{-\sigma} (\cos\theta') e^{-i\sigma\emptyset'} r^2 P_2^\sigma (\cos\theta) e^{i\sigma\emptyset}$$

is convenient in an M_S, M_J representation because only the
term with $\sigma = 0$ has a diagonal element (the non-vanishing
matrix elements have $M_S' = M_S - \sigma$, $M_J' = M_J + \sigma$). For the diagonal
element

$$(1/2)[3(\underline{S} \cdot \underline{r})^2 - \underline{S}^2 r^2] = (1/4)[3S_z^2 - \underline{S}^2][3z^2 - r^2]$$
$$= (1/4)[3M_S^2 - S(S+1)][3z^2 - r^2]. \qquad (6.04)$$

Thus for the state $S=1, M_S=1$; $J=1, M_J=1$ the matrix element
of the spin-spin interaction is
$$-\mu_P^2 \langle 1/r^3 \rangle \int (3\cos^2\theta - 1) |Y_1^1|^2 d\Omega,$$

with Y_1^1 the normalized angular factor of the orbital wave
function
$$Y_1^1 = -\sqrt{(3/8\pi)} P_1^1 (\cos\theta) e^{i\emptyset} = -\sqrt{(3/8\pi)} \sin\theta e^{i\emptyset}.$$
The angular integral is easily done, and equals $-2/5$, so we
find
$$\langle V_{spin-spin} \rangle_{M_S=1, M_J=1} = (2/5) \mu_P^2 \langle 1/r^3 \rangle. \qquad (6.05)$$

The M_J dependence of (6.04) can most easily be
obtained by using the theorem that the M dependence of the
matrix element is determined completely by the rotational
property of the operator: if two operators rotate in the

same way their matrix elements are proportional. Thus

$$(r^n Y_n^\sigma(\theta,\emptyset))_{J,M_J^\sigma;J,M_J} = K(J^n Y_n^\sigma(\theta,\emptyset))_{J,M_J^\sigma;J,M_J}, \quad (6.06)$$

where Y_n^σ is the normalized spherical harmonic and K is independent of σ and M_J, and for the diagonal element ($\sigma=0$)

$$(3z^2-r^2)_{J,M_J;J,M_J} = K[3M_J^2-J(J+1)].$$

The factor in the bracket on the right is unity for J=1, M_J=1, so (.05$) can be generalized to

$$\langle V_{spin-spin}\rangle_{M_S,M_J} = (2/5)\mu_P^2 \langle 1/r^3\rangle(3M_S^2-2)(3M_J^2-2), \quad (6.07)$$

which is required result for the strong field case.

     In the weak-field case $\underline{S}$ and $\underline{J}$ are coupled to a resultant $\underline{F}$ and, since $V_{spin-spin}$ is a rotational invariant, its matrix elements depend only on F and are independent of M_F. Their F dependence can be determined from a knowledge of the diagonal matrix elements in an M_S, M_J representation by a simple argument. We write

$$V_{spin-spin} = (2/5)\mu_P^2 \langle 1/r^3\rangle w. \quad (6.08)$$

In the M_S,M_J representation

$$w_{M_S M_J} = (3M_S^2-2)(3M_J^2-2).$$

In Table 6.1 the states are arranged according to M_F values and the values of $w_{M_S M_J}$ are given and also the trace of w, i.e. the sum of the diagonal elements, for each M_F value. If we transform from the M_S,M_J representation to the F,M_F representation, for a given M_F we obtain once each state with F≥M_F. Since the trace is invariant under the

TABLE 6.1						
M_F	M_S	M_J	$w_{M_S M_J}$	TR $w_{M_S M_J}$	w_F	
2	1	1	1	1	1	F=2
1	1	0	-2	-4	-5	F=1
	0	1	-2			
	1	-1	1			
0	-1	1	1	6	10	F=0
	0	0	4			

transformation
$$(Tr\ w)_{M_F} = \sum_{F=M_F}^{S+J} w_F.$$
The value of w_F is then given by
$$w_F = (Tr\ w)_{M_F=F} - (Tr\ w)_{M_F=F+1}.$$
The values of w_F are given in Table 6.1, and (6.08) gives
the spin-spin interaction in the weak-field case.

The relation (6.06) suggests that by using (6.03) we
could get an explicit expression for the F dependence of w
as proportional to $S^2J^2P_2(\cos\theta_{SJ})$. One is tempted to write
this as proportional to
$$3(\underline{S}\cdot\underline{J})^2 - \underline{S}^2\underline{J}^2,$$
but this is incorrect. Care must be taken with (6.03) in
the case when both the U_i and V_j are non commuting
operators. We suppose, however, that the U_i commute with
the V_j. We can see the trouble if we write out $(\underline{U}\cdot\underline{V})^2$ in
terms of the components
$$(\underline{U}\cdot\underline{V})^2 = \Sigma_{ij} U_iU_jV_iV_j = (1/4)\Sigma_{ij}(U_iU_j+U_jU_i)(V_iV_j+V_jV_i)$$
$$+(1/4)\Sigma_{ij}(U_iU_j-U_jU_i)(V_iV_j-V_jV_i).$$
The first term on the right is the contracted product of
symmetric tensors. The second term is
$$(1/2)(\underline{U}\times\underline{U})\cdot(\underline{V}\times\underline{V}),$$
the product of the vectors which can be constructed from
U_iU_j and V_iV_j if the components do not commute. If (6.03)
is to be true this unwanted term must be subtracted from
$(\underline{U}\cdot\underline{V})^2$, so
$$U^2V^2P_2(\cos\theta_{UV}) = (1/2)[3(\underline{U}\cdot\underline{V})^2 - (3/2)(\underline{U}\times\underline{U})\cdot(\underline{V}\times\underline{V}) - \underline{U}^2\underline{V}^2].$$
One learns in courses on quantum mechanics that a classical
quantity does not uniquely define a quantum operator; it is
perhaps surprising how rarely one must consciously take
this into account. In the classical limit the correction
term is of order h^2 (returning to c.g.s. units) compared to
the leading term, since each commutator is proportional to
h.

In our case $\underline{U}=\underline{S}$, $\underline{V}=\underline{J}$, and $\underline{S}\times\underline{S}=i\underline{S}$, $\underline{J}\times\underline{J}=i\underline{J}$. We can thus write

$$P_2(\cos\theta_{SJ}) = \frac{1}{2} \frac{[3S\cdot J(S\cdot J+1/2)-S(S+1)J(J+1)]}{S(S-1/2)J(J-1/2)}. \qquad (6.09)$$

The F dependence of $(\underline{S}\cdot\underline{J})$ is given by

$$\underline{S}\cdot\underline{J}=(1/2)[F(F+1)-S(S+1)-J(J+1)]$$

In (6.09) the proportionality factor has been chosen so that in the state F=S+J we have $P_2=1$. This choice is inappropriate if either S or J is 0 or 1/2. In either case the numerator is zero, as it should be, since an operator rotating as P_2 can have no diagonal matrix element in a state of angular momentum 0 or 1/2. The vanishing of the numerator if S or J is 1/2 can be used as an independent argument for the necessity of adding the term $(3/2)\underline{S}\cdot\underline{J}$.

With the above definition $w=P_2(\cos\theta_{SJ})$. It can easily be checked that (6.09) duplicates the answers shown in Table 6.1.

In fact, the Rabi experiment was done under strong field conditions, with H≈1500 gauss, more than an order of magnitude larger than the intrinsic fields (the corrections are of order of the square of the ratio, and thus less than 1%). In this case the matrix elements of (6.02) are

$$E_{M_S M_J} = -2\mu_P H M_S - 2\mu_R H M_J - 2\mu_P H' M_S M_J + (2/5)\mu_P^2 \langle 1/r^3 \rangle (3M_S^2-2)(3M_J^2-2). \qquad (6.10)$$

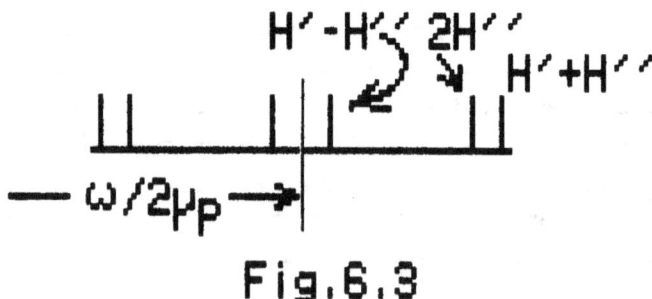

Fig.6.3

For the transitions with $\Delta M_S=\pm1$, (6.01) gives six
resonances at the fields, H, shown in Fig 6.3. We have
defined H'' to be $H''=(3/5)\mu_P\langle 1/r^3\rangle$.

The center of the pattern gave $\omega/2\mu_P$; this yielded the
same value of μ_P as the measurement for the J=0 state of
HD. The changes in magnetic field to pass from one
resonance to another yielded the values of H' and H'',
which were found to give

$$H'=27.2\pm0.3 \text{ gauss,}$$

$$\mu_P\langle 1/r^3\rangle=(5/3)H''=34.1\pm0.03 \text{ gauss.} \qquad (6.11)$$

Using the value $\langle 1/r^3\rangle=3.438\times10^{-24} \text{ cm}^{-3}$, calculated for the
H_2 molecule by Nordsieck, gave

$$\mu_P=2.785\pm0.03 \mu_0,$$

in agreement with the other results.

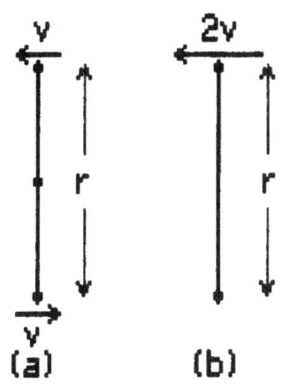

Fig.6.4

The calculation of H' is
not simple, but a rough
estimate will suffice to show
that the value obtained is
reasonable. In the laboratory
system, the magnetic field
produced by one proton at the
position of the other is (see
Fig 6.4 (a))

$$H'=ev/r^2,$$

or, since J=Mvr,

$$H'=eJ/Mr^3.$$

Comparing this with (6.11), which can be written

$$2.785(e/2M)\langle 1/r^3\rangle=34.1 \text{ gauss,}$$

we find

$$H'=(2/2.785)\times34.1=24.5 \text{ gauss.}$$

To this should be added the magnetic field produced by
rotation of the electrons, but even the sign of their
contribution is not clear: one can see that an electron
rotating between the protons will add to H', while a cloud

at a radius greater than $(1/2)r$ will subtract. The
experimental result indicates a small addition.

We have calculated the field in the laboratory system;
what is actually required is the field in a coordinate
system in which the proton is at rest. Under a Galilean
transformation to a system moving with velocity $\underline{v}$, the
magnetic field in the moving system is related to the
laboratory fields by

$$\underline{H}_v = \underline{H} - (\underline{v} \times \underline{E}).$$ (6.12)

However $\underline{E}=0$; the vanishing of the electric field is
precisely the condition which determines the equilibrium
position of the proton in the molecule[3]. Note that in the
comoving system the magnetic field produced by by the other
proton alone is $2H'$, as shown by Fig 6.4 (b), or by (6.12)
including the $\underline{E}$ field produced by the other proton.

---

[3] The small field required to balance the centrifugal force
for a rotating molecule can be seen to be of order of the
ratio of electron to proton masses smaller than the Coulomb
field between the protons. The effect of the Thomas
precession is also of this magnitude.

SEC.6II ELECTRIC QUADRUPOLE MOMENT OF THE DEUTERON

For the D_2 molecule the spin-spin magnetic interaction is

$$V_{spin-spin} = -(\mu_D^2/r^3)[3(\underline{S}_1 \cdot \underline{r})(\underline{S}_2 \cdot \underline{r}) - (\underline{S}_1 \cdot \underline{S}_2)r^2]/r^2.$$

The expression similar to the second form of (6.02) is

$$V_{spin-spin} = -(\mu_D/r^3) \cdot (1/2r^2)\{[3(\underline{S} \cdot \underline{r})^2 - \underline{S}^2 r^2] - [3(\underline{S}_1^2 \cdot \underline{r}) - \underline{S}_1^2 r^2]$$
$$- [3(\underline{S}_2^2 \cdot \underline{r}) - \underline{S}_2^2 r^2]\}, \qquad (6.13)$$

as can be checked by writing $\underline{S} = \underline{S}_1 + \underline{S}_2$ in $[3(\underline{S} \cdot \underline{r})^2 - \underline{S}^2 r^2]$ and multiplying out. This form has the advantage that each term involves only two vectors and is thus of the form (6.03).

The measurements were made on the J=1 state of D_2, which must have S=1. There are two possible $M_S=1$ states, that with $M_{S_1}=1$, $M_{S_2}=0$ and that with $M_{S_1}=0$, $M_{S_2}=1$. The S=1 state is the antisymmetric combination of these two. When the addition theorem (6.03) is used, $(3M_S^2-2)=1$ for this state, and the expectation values of $(3M_{S_1}^2-2)$ and $(3M_{S_2}^2-2)$ are each $-1/2$ and so double the contribution of the first term in (6.13). Thus the interaction is effectively

$$V_{spin-spin} = -(\mu_D^2/r^3)[3(\underline{S} \cdot \underline{r})^2 - \underline{S}^2 r^2]/r^2. \qquad (6.14)$$

This differs from (6.02) by a factor 2. However, the first term in (6.02) would now be $-\mu_D \underline{S} \cdot \underline{H}$, also missing a factor 2 (because of the different g-factor for spins 1 and 1/2), so the factor 2 drops out in calculating H'', and we have

$$H'' = (3/5)\mu_D \langle 1/r^3 \rangle.$$

The measurements gave H'=14 gauss, which agrees with the expectation that H' is inversely proportional to mass. The value of $\mu_D \langle 1/r^3 \rangle$ would be expected to be $(\mu_D/\mu_P) \times 34.1 = 10.5$ gauss. However, the measured value was 98.1 gauss. The conclusion must be that there is an additional interaction with similar rotational behavior, i.e. one that gives an energy proportional to $P_2(\cos\theta_{SJ})$. The only physically plausible candidate is is an electric quadrupole interaction, which would require the deuteron to have a quadrupole moment. Thus the deuteron can not be in a pure spherically symmetric s state; some fraction of higher

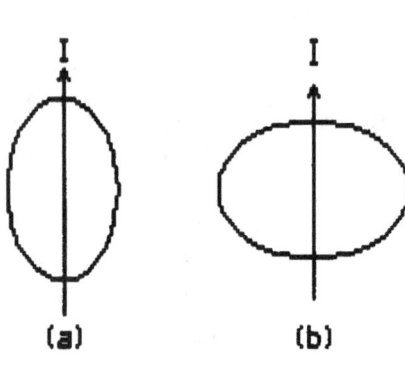

(a)   (b)

Fig.6.5

l must be present to give a non-symmetric charge distribution in the ground state, such as shown in Fig 6.5 (a) or (b). We have denoted the total angular momentum of the deuteron by $\underline{I}$ (I=1), distinguishing it from $\underline{S}$, which will no longer be a constant of the motion. Since there is no preferred direction other than $\underline{I}$, the

charge distribution will have axial symmetry about $\underline{I}$. We are concerned with the interaction of this charge density with the charge density of the remainder of the molecule, which, similarly, will have axial symmetry about $\underline{J}$. This, then, will lead to a term in the energy with the right dependence on rotations.

The Coulomb energy of one of the deuterons, call it deuteron 1, is

$$V_{1,Coul} = \int \rho_1(\underline{r}) \rho'(\underline{r}') / |\underline{r}-\underline{r}'| \, d\underline{r} d\underline{r}' ,$$

where $\rho_1(\underline{r})$ is the deuteron charge density and $\rho'(\underline{r}')$ is the charge density of the remaining charges in the molecule. If we take the origin of coordinates at the center of gravity of the deuteron and expand $1/|\underline{r}-\underline{r}'|$ in spherical harmonics we obtain

$$V_{1,Coul} = \int \rho_1(\underline{r}) \rho'(\underline{r}') (r^n/r'^{n+1}) P_n(\cos\theta_{rr'}) \, d\underline{r} d\underline{r}' ,$$

if we neglect the small contribution from $r' < r$. The monopole term, n=0, is taken into account in solving the molecular binding problem. The n=1 term is the electric dipole contribution, but this vanishes, since the

expectation value of the deuteron dipole moment is zero if
parity is conserved. The n=2 term gives the quadrupole
interaction. If we use the addition theorem, (6.03), in the
strong field case with the z axis in the direction of the
external magnetic field, we get (omitting the $\sigma \neq 0$ terms,
which contribute no diagonal elements in an M_{I1}, M_{I2}, M_J
representation)

$$V_{1,Quad} = er_P^2 P_2(\cos\theta_{r_Pz}) \cdot \Sigma_i e_i P_2(\cos\theta_{r_iz})/r_i^3,$$

$$= (1/4)e(3z_P^2 - r_P^2) \cdot \Sigma_i (e_i/r_i^3)(3z_i^2 - r_i^2)/r_i^2, \qquad (6.15)$$

where $\underline{r}_P$ is the coordinate of the proton in the deuteron
and the index i refers to the other charges in the molecule.

Let us define quantities q and Q by

$$eq = \langle \Sigma_i (e_i/r_i^3)(3z_i^2 - r_i^2)/r_i^2 \rangle_{M_J=1},$$

$$eQ = e\langle 3z_P^2 - r_P^2 \rangle_{M_{I1}=1}.$$

It can easily be checked that

$$eq = \langle -\partial E_z/\partial z \rangle_{M_J=1},$$

where E_z is the z component of the electric field at the
position of the deuteron; eQ is the electric quadrupole
moment of the deuteron. Q itself is commonly called the
quadrupole moment of the deuteron. It follows from the
theorem (6.06) that the diagonal matrix elements in the
M_{I1}, M_{I2}, M_J representation are

$$\langle V_{quad} \rangle_{M_{I1}, M_{I2}, M_J} = (1/4)e^2 qQ[(3M_{I1}^2-2)+(3M_{I2}^2-2)](3M_J^2-2).$$

If we change from the M_{I1}, M_{I2} representation to the I,M_I
representation ($\underline{I}=\underline{I}_1+\underline{I}_2$), it follows from the discussion of
(6.13) that in the I=1 state

$$\langle V_{quad} \rangle_{M_I, M_J} = -(1/4)e^2 qQ[(3M_I^2-2)(3M_J^2-2).$$

Adding the spin-spin interaction we have (see (6.10), and
remember the factor 2 in (6.14))

$$\langle V_{spin-spin} + V_{quad} \rangle_{M_I M_J}$$

$$= [(1/5)\mu_D^2 \langle 1/r^3 \rangle - (1/4)e^2 qQ](3M_I^2-2)(3M_J^2-2)$$

or, more generally for the I=1,J=1 states,

$$V_{spin-spin} + V_{quad} = [(1/5)\mu_D^2 \langle 1/r^3 \rangle - (1/4)e^2 qQ]P_2(\cos\theta_{IJ}).$$

The measured value, $(5/3)H'' = 98.1$ gauss, thus is not an evaluation of $\mu_D \langle 1/r^3 \rangle$ but of

$$\mu_D \langle 1/r^3 \rangle - (5/4)e^2 qQ/\mu_D.$$

To find Q, q must be calculated from the molecular wave function; the final result is

$$Q = 2.73 \times 10^{-27} cm^2.$$

The positive sign of Q means that the shape resembles Fig 6.5 (a). However the distortion is not large: its measure is

$$\frac{Q}{\langle r_P^2 \rangle} = 3 \frac{\langle z_P^2 \rangle}{\langle r_P^2 \rangle} - 1 = 0.07.$$

Here we have evaluated $\langle r_P^2 \rangle$ using the approximate deuteron wave function given on page 68, with the result

$$\langle r_P^2 \rangle = (1/4)\langle r^2 \rangle = 1/[8(1-r_t\beta)\beta^2] = 3.94 \times 10^{-26} cm^2.$$

Measurements of the spectrum of HD in the $J=1$ state also agreed with the interpretation of the anomaly as due to a deuteron electric quadrupole moment, and gave the same value of Q.

SEC.6III THE TENSOR FORCE

If the ground state of the deuteron is slightly non-spherical it means that a small amount of states of $l>0$ are included in its wave function. Conservation of parity requires that only states of even l are in question. The requirement that the total angular momentum, which of course remains a constant of the motion, is $I=1$ reduces the choice to a single state: the $^3D_1$ state alone has the same parity and angular momentum as the $^3S_1$ state.

One might attempt to estimate the percentage of d state in the ground state of the deuteron by the slight departure of the deuteron magnetic moment from additivity:

$$\mu_P=2.79278\pm0.00002,$$
$$\mu_N=-1.91315\pm0.00007,$$
$$\mu_N+\mu_P=0.87953,$$
$$\mu_D =0.85742,$$
$$\mu_N+\mu_P-\mu_D=0.02211.$$

Suppose that the deuteron moment is given (in nuclear magnetons) by

$$\mu_D=\mu_P\underline{\sigma}_P+\mu_N\underline{\sigma}_N+(1/2)\underline{L},$$

the last term representing the orbital magnetic moment (the factor 1/2 because only the proton is charged). This can be written

$$\mu_D=(1/2)(\mu_P+\mu_N)(\underline{\sigma}_P+\underline{\sigma}_N)+(1/2)(\mu_P-\mu_N)(\underline{\sigma}_P-\underline{\sigma}_N)+(1/2)\underline{L}$$
$$=(\mu_P+\mu_N)\underline{S}+(1/2)\underline{L}+(1/2)(\mu_P-\mu_N)(\underline{\sigma}_P-\underline{\sigma}_N).$$

The last term has no diagonal matrix element (it gives the triplet-singlet transitions responsible for the magnetic dipole photoeffect) and can be dropped. Since $\underline{I}=\underline{S}+\underline{L}$, we can write

$$\mu_D=(\mu_P+\mu_N)\underline{I}-(\mu_P+\mu_N-1/2)\underline{L}.$$

Using the usual vector model, we have for the z component

$$\mu_{Dz}=(\mu_P+\mu_N)I_z-(\mu_P+\mu_N-1/2)\frac{\underline{L}\cdot\underline{I}I_z}{I(I+1)}$$
$$=\{(\mu_P+\mu_N)-(\mu_P+\mu_N-1/2)\cdot(1/2)\frac{[I(I+1)+L(L+1)-S(S+1)]}{I(I+1)}\}I_z.$$

The deuteron magnetic moment is the expectation value of

this operator in the $I=S=1$, $I_z=1$ ground state, so
$$\mu_D=\mu_P+\mu_N-(1/4)(\mu_P+\mu_N-1/2)\langle L(L+1)\rangle,$$
the last term representing the departure from additivity.
If a fraction p_d of d state is included in the deuteron
wave function
$$\langle L(L+1)\rangle=2\cdot 3p_d=6p_d,$$
and
$$\mu_D=\mu_P+\mu_N-(3/2)(\mu_P+\mu_N-1/2)p_d.$$
This gives
$$p_d=\frac{2}{3}\frac{(\mu_P+\mu_N-\mu_D)}{(\mu_P+\mu_N-1/2)}=\frac{2}{3}\frac{(0.02211)}{(0.37953)}=0.039.$$

However this estimate cannot be considered at all
reliable, since, as already remarked in connection with the
N-P capture cross section, the assumption that the free
proton and neutron magnetic moments can be used in the
expression for the deuteron moment, and the contribution of
exchange currents neglected, can lead to errors of a few
percent. Calculations of the deuteron wave function using
the best information on the N-P forces, which, of course,
are chosen to fit the observed quadrupole moment, give
values more like $p_d=0.07$.

If we now ask what kind of interaction between proton
and neutron can couple the ${}^3S_1$ and ${}^3D_1$ states we see we
must have an interaction which behaves like $P_2(\cos\theta)$ under
rotations of r, like the last term of (6.02). If we
restrict ourselves to velocity independent forces, the only
rotational invariant we con construct which has this
property, for spin 1/2 particles, can differ from that
magnetic interaction only in its dependence on r:
$$V_{tensor}=V(r)[3(\underline{\sigma}_P\cdot\underline{r})(\underline{\sigma}_N\cdot\underline{r})-(\underline{\sigma}_P\cdot\underline{\sigma}_N)r^2]/r^2, \qquad (6.16)$$
the so-called tensor force.

An interaction of this form can be obtained from meson
theory if we change the meson field equation, (1.06), by
coupling the meson field to the spin density $\underline{\sigma}\rho$, rather
than to ρ:

$$\Delta\emptyset - \mu^2\emptyset = -(1/\mu)\,\text{div}\,\underline{g}\rho. \tag{6.17}$$

The div is required if $\emptyset$ is to rotate as a scalar. However, under reflections, because of the extra spatial derivative, $\emptyset$ behaves as a pseudoscalar, $\emptyset(\underline{r}) \to -\emptyset(-\underline{r})$. Then, in place of the scalar interaction, (1.04),

$$V = -(1/8\pi)\int (e^{-\mu|\underline{r}'-\underline{r}''|}/|\underline{r}'-\underline{r}''|)\rho(\underline{r}')\rho(\underline{r}'')\,d\underline{r}'d\underline{r}'',$$

which, for point neutron and proton separated by a distance $\underline{r}=\underline{r}_P-\underline{r}_N$, $\rho(\underline{r}')=g_P\delta(\underline{r}'-\underline{r}_P)+g_N\delta(\underline{r}'-\underline{r}_N)$, gives the Yukawa interaction between the two particles,

$$V_{NP} = -(g_Pg_N/4\pi)e^{-\mu r}/r,$$

we have

$$V = -(1/8\pi\mu^2)\int \frac{e^{-\mu|\underline{r}'-\underline{r}''|}}{|\underline{r}'-\underline{r}''|}\,\text{div}'\,\underline{g}\rho(\underline{r}')\,\text{div}''\,\underline{g}\rho(\underline{r}'')\,d\underline{r}'d\underline{r}'',$$

or, by partial integration,

$$V = (1/8\pi\mu^2)\int [\rho(\underline{r}')\underline{g}\cdot\text{grad}'][\rho(\underline{r}'')\underline{g}\cdot\text{grad}']\frac{e^{-\mu|\underline{r}-\underline{r}''|}}{|\underline{r}-\underline{r}''|}\,d\underline{r}'d\underline{r}'',$$

and, for point N and P,

$$V_{NP} = (g_Pg_N/4\pi\mu^2)(\underline{g}_P\cdot\text{grad})(\underline{g}_N\cdot\text{grad})e^{-\mu r}/r.$$

This V is the sum of a tensor force and a spin dependent spherical force, $V=V_{tensor}+V_{spherical}$, with

$$V_{tensor} = (g_Pg_N/4\pi\mu^2)[(\underline{g}_P\cdot\text{grad})(\underline{g}_N\cdot\text{grad})-(1/3)(\underline{g}_P\cdot\underline{g}_N)\Delta]e^{-\mu r}/r$$

$$= (g_Pg_N/12\pi\mu^2)[1+(3/\mu r)+(3/\mu^2 r^2)]e^{-\mu r}/r$$

$$\cdot[3(\underline{g}_P\cdot\underline{r})(\underline{g}_N\cdot\underline{r})-(\underline{g}_P\cdot\underline{g}_N)r^2]/r^2, \tag{6.18}$$

$$V_{spherical} = (g_Pg_N/12\pi\mu^2)\underline{g}_P\cdot\underline{g}_N\Delta e^{-\mu r}/r$$

$$= (g_Pg_N/12\pi)\underline{g}_P\cdot\underline{g}_N[e^{-\mu r}/r + (1/\mu^2)\delta(\underline{r})] \tag{6.19}$$

Note that a positive Q requires that V_{tensor} be attractive for $\underline{r}$ parallel to $\underline{S}$, so g_P and g_N must have opposite signs. The like particle forces are given by the same formula with g_Pg_N replaced by g_P^2 or g_N^2.

This pseudoscalar meson theory represents the interaction of the nucleons with π mesons. Since the π is the lightest meson, and therefore gives the interaction with the longest range, we may expect that (6.18) and (6.19) represent (aside from the question of exchange forces, i.e. of the contributions of charged as well as neutral mesons) the forces at large distances.

As shown in (6.02), the interaction (6.16) can also be written

$$V_{tensor} = 2V(r)[3(\underline{S} \cdot \underline{r})^2 - \underline{S}^2 r^2]/r^2, \qquad (6.20)$$

and thus, even with the inclusion of the tensor force, $\underline{S}^2$ (though not the components S_x, S_y, S_z) remains a constant of the motion. In fact, it follows from (6.20) that the tensor force vanishes for the singlet states and only exists for the triplets. If we list the even parity triplet states we have

$$^3S_1; {}^3D_{3,2,1}; {}^3G_{5,4,3}; \cdots,$$

and it follows, since total angular momentum must be conserved, that the only states that can combine are

$$^3S_1 + {}^3D_1; {}^3D_2; {}^3D_3 + {}^3G_3; \cdots.$$

Similarly, for the odd parity triplet states we have

$$^3P_{2;1,0}; {}^3F_{4,3,2}; {}^3H_{6,5,4}; \cdots,$$

and the possible combinations are

$$^3P_0; {}^3P_1; {}^3P_2 + {}^3F_2; {}^3F_3; {}^3F_4 + {}^3H_4; \cdots.$$

The singlet states

$$^1S_0; {}^1D_2; {}^1G_4; {}^1P_1; {}^1F_3; {}^1H_5; \cdots$$

all have different total angular momentum, and cannot combine.

We may wonder why S is conserved, e.g. why not $^3D_2 + {}^1D_2$? The spin selection rule will be true for any Hamiltonian which is invariant under interchange of proton and neutron coordinates (space and spin), so that the exchange operator $P_{PN} = P_{PN}^{(x)} P_{PN}^{(\sigma)}$ is a constant of the motion (this requires neglect of the N-P mass difference). Since, for the deuteron internal wave function, the space exchange operator is the same as the parity operator, each changing $\underline{r}$ to $-\underline{r}$, conservation of parity means conservation of $P_{PN}^{(x)}$. The conservation of P_{PN} and $P_{PN}^{(x)}$ implies the conservation of $P_{PN}^{(\sigma)}$. Thus the triplet states, which are even under spin exchange $(P_{PN}^{(\sigma)}=1)$, cannot mix with the odd sinlet states $(P_{PN}^{(\sigma)}=-1)$.

The $^3D_1$ wave functions are of the form

$$^3D_{1,M_I} = \frac{w(r)}{r} N \frac{[3(S\cdot r)^2-S^2 r^2]}{r^2} \chi^1_{M_I},$$

where χ^S_M is the spin function, and N is chosen to normalize the wave function with respect to the angular integration and spin sum; this because it is evident that these wave functions rotate in the same way as the spin function, and thus have total angular momentum I=1 and $I_z=M_I$, while also being second spherical harmonics in $\underline{r}$ and S=1 in spin. Thus, if a tensor force is present, in place of the usual triplet s wave function,

$$\psi_{M_S}=[u(r)/r]\chi^1_{M_S},$$

we have

$$\psi_{M_I} = \{\frac{u(r)}{r} + \frac{w(r)}{r} N \frac{[3(S\cdot r)^2-S^2 r^2]}{r^2}\}\chi^1_{M_I}. \qquad (6.21)$$

Substitution of this form in the Schrodinger equation leads to coupled equations for u and w,

$$-\frac{1}{M} u'' +[V_c(r)-E]u+2\sqrt{2}V_t(r)w=0,$$
$$-\frac{1}{M} w'' +[V_c(r)-2V_t(r)+\frac{6}{Mr^2} -E]w+2\sqrt{2}V_t(r)u=0,$$

where $V_c(r)$ is whatever central force is present, and $V_t(r)$ is the V(r) of (6.16).

If we put $E=-\epsilon_D=-\beta^2/M$, and measure distances in units of the deuteron radius, i.e. write $r'=\beta r$, the equations become

$$d^2u/dr'^2=-(U+1)u-2\sqrt{2}Ww=0,$$
$$d^2w/dr'^2=-(U-2W+\frac{6}{r'^2} +1)w-2\sqrt{2}Wu=0, \qquad (6.22)$$

with

$$U=V_c/\epsilon_D, \quad W=V_t/\epsilon_D.$$

In these units the observed quadrupole moment is

$$Q'=Q\beta^2=0.0147.$$

Despite the smallness of $4Q'$ (the factor 4 because Q was defined in terms of $r_p^2 = r^2/4$), the required magnitude of the tensor force, in the case where its range is small compared to $1/\beta$, cannot be much less than that of the central forces required to bind the deuteron in the central force case discussed in Chapter 2. That is, V_t must also be of order $\pi^2/(4MR^2)$ (see (2.04)). To see this, let us consider the simplest case, where V_c and V_t are square well potentials of range R, with $R' = \beta R \ll 1$.

Near $r' = 0$, $u = ar'$, $w = br'^3 \ln r'$. For $r' > R'$, U and W are zero, and the solutions of (6.22) with the proper behavior as $r' \to \infty$ are

$$u \approx \sqrt{2} e^{-r'}, \qquad w \approx N' e^{-r'} [1 + 3/r' + 3/r'^2]. \qquad (6.23)$$

(Note that $u \approx c_0 \sqrt{r'} H_{1/2}^{(1)}(ir')$, $w \approx c_2 \sqrt{r'} H_{5/2}^{(1)}(ir')$, $H_n^{(1)}$ being the Hankel function. The solutions of the free particle equation, it should be remembered, are $u_1 = c_1 \sqrt{(kr)} J_{1+1/2}(kr)$.) The normalization of u is chosen to be correct in the zero-range approximation. The solutions thus must look something like Fig 6.6. We suppose $w \ll u$.

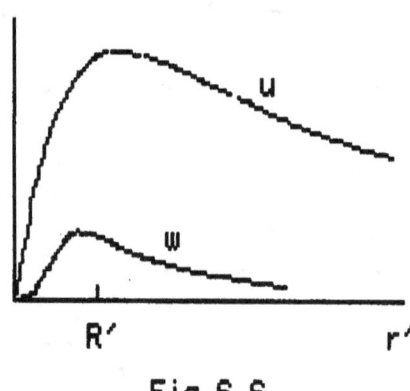

R′   r′

Fig.6.6

If the quadrupole moment is calculated from the wave function (6.21) it is found that (neglecting a w^2 term)

$$Q' = 1/(5\sqrt{2}) \int_0^\infty uwr'^2 dr'.$$

Because of the r'^2 weighting, the contributions to the integral come from $r' > R'$, so we may use (6.23) for u and w, which gives

$$Q' = (N'/5) \int_0^\infty e^{-2r'} (r'^2 + 3r' + 3) dr' = (1/2) N'. \qquad (6.24)$$

If $w<u$ we can deduce from (6.24) a minimum range for
the tensor force. We have, to leading order in $1/R'$,
$$w(R')=(3N'/R'^2)<u(R')=\sqrt{2};$$
thus
$$R'^2>3N'/\sqrt{2}=3\sqrt{2}Q', \quad \text{or } R^2>3\sqrt{2}Q.$$
Using the observed Q', we find $R'>0.25$, or $R>1.1$ f. (The
assumption $R'<<1$ is not too well satisfied, to about the
same extent that our previous zero-range approximation was
inadequate.)

The d wave probability is
$$P_d=\int_0^\infty w^2 dr'.$$
Without going into the question of the contribution to the
integral from $r'<R'$, we can write
$$P_d>\int_{R'}^\infty w^2 dr',$$
or, to leading order in $1/R'$,
$$P_d>\int_{R'}^\infty (9N'^2/r'^4)dr'=(3N'^2/R'^3)=12Q'^2/R'^3=12Q^2\beta/R^3.$$
For $R=1.7$ f, the triplet effective range, the quantity on
the right equals 0.04, which is thus a reasonable magnitude
for P_d.

We can obtain an estimate of the required magnitude of
the tensor force by multiplying the first of equations
(6.22) by w, the second by u, and subtracting, which gives
$$dT(u,w)/dr'+[(6/r'^2)-2W]uw+(2\sqrt{2})W(u^2-w^2)=0,$$
with $T=uw'-wu'$. If this relationship is integrated between
0 and ∞ we have, since T vanishes at both limits,
$$2\int_0^\infty(-W)(\sqrt{2}u^2-2uw-\sqrt{2}w^2)dr'=6\int_0^\infty uw/r'^2 dr'.$$
The uw and w^2 terms on the left are presumably small
compared to the u^2 term, but even without this supposition
we have
$$2\sqrt{2}\int_0^\infty(-W)u^2 dr'>6\int_0^\infty uw/r'^2 dr'.$$
We may suppose that in the well u is not too different from

what it would be for the central force problem,
$u = \sqrt{2}\sin(\pi r'/2R')$, so

$$\int_0^{R'} u^2 dr' = R',$$

from which it follows that

$$-WR' > (3/\sqrt{2})\int_{R'}^{\infty} uw/r'^2 dr' = 9N'\int_{R'}^{\infty} dr'/r'^4 = 3N'/R'^3,$$

or

$$-W > 3N'R'^4 = 6Q'/R'^4.$$

In terms of the unscaled quantities this can be written

$$-V_t > (24Q/\pi^2 R^2)(\pi^2/4MR^2),$$

which, for $R = 1.7$ f, gives

$$-V_t > 0.23(\pi^2/4MR^2).$$

Thus, for a reasonable range, the magnitude of V_t cannot
be much smaller than the characteristic energy of the
nuclear forces.

SEC.6IV NEUTRON MAGNETIC MOMENT

   The neutron magnetic moment can be measured by the
magnetic resonance method if a polarizing and analyzing
device is available to replace the A and B magnets of
Fig 6.1. Alvarez and Bloch[*], who did the first experiments,
replaced the A and B magnets by magnetized iron plates
(Fig.6.7). As we saw in our discussion of neutron

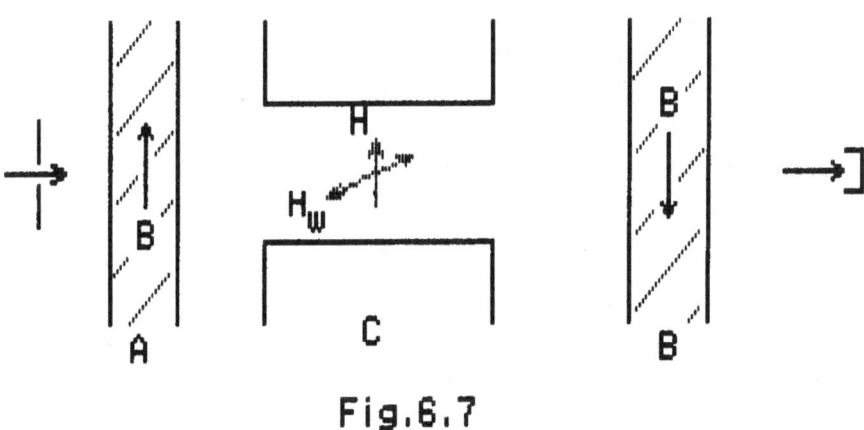

Fig.6.7

reflection from ferromagnetic mirrors (Sec.3V), the
forward scattering amplitude due to the magnetic
interaction is comparable to that of nuclear scattering,
which implies that the total scattering cross section can
be appreciably different for neutrons with spins parallel
or antiparallel to the magnetic field. The attenuation of
the neutron beam produced by an iron plate will thus be
different for the two directions of neutron spin. Let f_p
and f_a be the fraction of unscattered neutrons emerging

---

[*] L.A. Alvarez and F. Bloch, Phys. Rev. <u>57</u>, 111 (1940)

from a plate for spin parallel or antiparallel to the
field. For an unpolarized initial beam, the fraction
reaching the detector with the arrangement of Fig 6.7 will
be $f_p f_a$ if the spins are unchanged in the C region, but
$(1/2)(f_p^2+f_a^2)$ if the spins are flipped. Since
$(1/2)(f_p^2+f_a^2) > f_p f_a$ if $f_p \neq f_a$, the detected intensity will
increase at resonance. Alvarez and Bloch found
$\mu_N = 1.93 \pm 0.02$.

The dependence of the scattering cross section of an
iron atom on the spin direction is readily calculated. The
interaction of the neutron magnetic moment with the
magnetic field, $\underline{b}$, produced by the electrons is

$$V = -(e/2M)\mu_N \underline{\sigma} \cdot \underline{b},$$

and, in Born approximation, this leads to a scattering
amplitude for scattering with momentum change $\underline{q}$,

$$f_{\underline{q}} = -a - (M/2\pi)V_{\underline{q}} = -a + (e/4\pi)\mu_N \underline{\sigma} \cdot \underline{b}_{\underline{q}}, \qquad (6.25)$$

with a the scattering amplitude of the nucleus.

The Fourier component $\underline{b}_{\underline{q}}$ is most easily calculated by
remembering the equation satisfied by the vector potential,

$$\Delta \underline{A} = -4\pi \underline{j},$$

which gives

$$\underline{A}_{\underline{q}} = 4\pi \underline{j}_{\underline{q}}/q^2.$$

We have, since $\underline{b} = \text{curl } \underline{A}$,

$$\underline{b}_{\underline{q}} = i\underline{q} \times \underline{A}_{\underline{q}}$$

and, if $\underline{m}(\underline{r})$ is the magnetic moment density of the
electrons, $\underline{j} = \text{curl } \underline{m}$ or

$$\underline{j}_{\underline{q}} = i\underline{q} \times \underline{m}_{\underline{q}}.$$

Thus

$$\underline{b}_{\underline{q}} = -4\pi \underline{q} \times (\underline{q} \times \underline{m}_{\underline{q}})/q^2 = 4\pi[\underline{m}_{\underline{q}} - (\underline{q} \cdot \underline{m}_{\underline{q}})\underline{q}/q^2] \qquad (6.26)$$

The expression (6.26) has a peculiarity, that its
limit as $\underline{q} \to 0$ depends on the relative directions of $\underline{m}_{\underline{q}}$ and
$\underline{q}$, which has sometimes led to confusion in applying the
optical theorem to obtain the mean potential in a
magnetized medium, despite the fact that by definition the

mean field in the medium is the macroscopic field $\underline{B}$. The
reason for the apparent ambiguity is the infinite range of
the dipole field: with a $1/r^3$ field the distant portions of
the magnet contribute to the field at any point, and one
must pay attention to the actual configuration of the
magnet.

Let us examine the relationship between the
microscopic field $\underline{b}$ and the macroscopic field $\underline{B}$. The
Fourier component of $\underline{m}$ is the sum of the contributions of
all the atoms in the magnet, with the proper phase factors,

$$\underline{m}_{\underline{g}} = \Sigma_i \underline{m}_{i\underline{g}} e^{-i\underline{g}\cdot\underline{r}_i}, \qquad (6.27)$$

the index i running over all atoms and $\underline{r}_i$ being being the
position of the ith atom. The macroscopic field at a point
$\underline{r}$ is obtained by averaging the microscopic field over a
small volume $d\underline{r}$ around $\underline{r}$, but one large enough to include
many atoms. The contributions of the atoms in $d\underline{r}$ to the sum
is

$$\underset{\underline{r}_i \text{ in } d\underline{r}}{\Sigma} \underline{m}_{i\underline{g}} e^{-i\underline{g}\cdot\underline{r}_i} = (\underset{\underline{r}_i \text{ in } d\underline{r}}{\Sigma} \underline{m}_{i\underline{g}} e^{-i\underline{g}\cdot(\underline{r}_i-\underline{r})}) e^{-i\underline{g}\cdot\underline{r}}.$$

The effect of averaging the field over $d\underline{r}$ is to replace the
factor in the bracket by its value for $\underline{g}=0$ (note that this
does not change the value when $\underline{g}=0$; averaging changes $\underline{b}$ to
$\underline{B}$, with $\underline{b}_0=\underline{B}_0$), which is

$$\underset{\underline{r}_i \text{ in } d\underline{r}}{\Sigma} \underline{m}_{i0}.$$

Now

$$\underline{m}_{i0} = \int \underline{m}_i (\underline{r}) d\underline{r} = \underline{m}_i ,$$

the magnetic moment of the atom, and

$$\underset{\underline{r}_i \text{ in } d\underline{r}}{\Sigma} \underline{m}_i = N_i \underline{m}_i d\underline{r} = \underline{M}(\underline{r}) d\underline{r},$$

where N_i is the number of atoms/cc and $\underline{M}(\underline{r})$ is the
macroscopic magnetization. The complete sum in (6.27)
becomes

$$\int \underline{M}(\underline{r}) e^{-i\underline{g}\cdot\underline{r}} d\underline{r} = \underline{M}_{\underline{g}}.$$

Thus in place of (6.26) we have for the macroscopic field

$$\underline{B}_q = 4\pi[\underline{M}_q - (\underline{q}\cdot\underline{M})\underline{q}/q^2] + \underline{H}_0(2\pi)^3\delta(\underline{q}). \qquad (6.28)$$

We have added a term which gives a field $\underline{H}_0$ uniform in space; $\underline{H}_0$ can be chosen to fit a desired boundary condition at infinity.

As examples, consider the situation shown in Fig 6.8 (a) and (b), an infinite plate magnetized parallel or perpendicular to the surface. The magnetization is constant

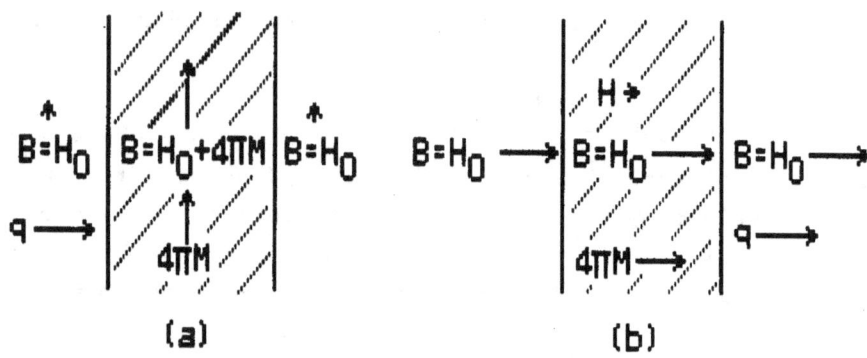

Fig.6.8

in the plate, and, since there is no variation in a direction parallel to the plate, $\underline{q}$ is perpendicular to the plate ($\underline{M}_q$ is the Fourier component of a step function ⎍⎍).

In case (a) $\underline{q}$ and $\underline{M}_q$ are perpendicular, $\underline{B}_q = 4\pi\underline{M}_q + \underline{H}_0(2\pi)^3\delta(\underline{q})$ and the Fourier inverse gives $\underline{B} = \underline{H}_0$ outside the plate, $\underline{B} = \underline{H}_0 + 4\pi\underline{M}$ inside. Since the transverse component of the macroscopic field $\underline{H}$ is continuous at the surface of the plate, in this case $\underline{H} = \underline{H}_0$ everywhere, and within the plate $\underline{B} = \underline{H} + 4\pi\underline{M}$, in agreement with the definition of $\underline{H}$. In case (b) $\underline{q}$ and $\underline{M}_q$ are parallel, $\underline{B}_q = \underline{H}_0(2\pi)^3\delta(\underline{q})$, and the Fourier inverse gives $\underline{B} = \underline{H}_0$ everywhere. This agrees with the boundary condition that the normal component of $\underline{B}$ is

continuous at the surface. In this case it is $\underline{H}$ which is
discontinuous, $\underline{H}=\underline{H}_0$ outside the plate, $\underline{H}=\underline{B}-4\pi\underline{M}$ within. The
reflection of neutrons by a magnetic mirror is carried out
with an arrangement approximating case (a). Since the
critical angle is determined by the change in potential
across the surface it depends, strictly speaking, on $4\pi\underline{M}$
rather than $\underline{B}$.

The scattering of neutrons in an iron plate is due to
the local variations in the magnetic field, i.e. to the
difference between the microscopic field $\underline{b}$ and the
macroscopic field $\underline{B}$. Since, for thermal neutrons,
$1/p=0.29$ A is considerably smaller than the mean
interatomic separation, 2.3 A, we may expect the absorption
coefficient is $N\sigma$, with σ the scattering cross section of
an individual atom. Using (6.25) and (6.26), we write the
scattering amplitude of an atom

$$f=-a-a_{mag}f(\underline{q})[\sigma_z-(\underline{\sigma}\cdot\underline{q})q_z/q^2],$$

where (see Sec.3V) $a_{mag}=(e^2/2m)n_B(-\mu_N)$, $f(\underline{q})$ is the form
factor of the magnetic moment distribution in the atom
($f(0)=1$), and the magnetic field is taken to be in the z
direction. If we set $f(\underline{q})=1$ we readily find for the cross
section

$$\sigma=4\pi\{a[a\pm(3/2)a_{mag}]+(3/4)a^2_{mag}\},$$

the $\pm$ sign referring to neutrons with spins parallel or
antiparallel to the magnetic field. With the values given
in Sec. 3V, $a=9.2$ f, $a_{mag}=5.58$ f, we find $\sigma=24.4$ b for
parallel spin, $\sigma=3.6$ b for antiparallel spin. The
difference is undoubtedly exaggerated by neglect of the
factor $f(\underline{q})$, particularly in light of the large
cancellation between a and $-(3/2)a_{mag}$ in the antiparallel
case, but the estimate suffices to show a considerable
difference in absorption coefficient can be expected.

A troublesome difficulty in the Bloch-Alvarez method
of polarizing neutrons is that nearly complete saturation

of the magnet is required. One must remember the domain
structure of a ferromagnet; if full saturation is not
achieved the magnetic field points in a somewhat different
direction in each small domain. Each time a neutron crosses
a domain surface to the next domain where the field has
changed direction by an angle θ it has a probability
P=sin²θ/2 of reversing its spin direction, the cumulative
effect leading to a rapid loss of polarization.

The highly precise value of μ_N quoted in Sec.6II was
obtained using the intense source of thermal neutrons from
the Brookhaven reactor. Hughes and Burgy[a] were able to to
use magnetized mirrors as the polarizer and analyzer. The
mirrors were a cobalt-iron alloy; adding a few percent of
iron makes it easier to reach the saturation field.

---

[a] D.J. Hughes and M.T. Burgy, Phys. Rev. 81 498 (1951).

SEC.6V NEUTRON-ELECTRON INTERACTION

   Beginning in 1947, Rabi and his collaborators at
Columbia and Fermi and his collaborators at Chicago[a]
independently conceived and carried out experiments with
thermal neutrons to measure an electrostatic interaction
between neutrons and electrons. The origin of such as
interaction can be readily imagined in terms of meson
theory: for a certain fraction of the time the neutron
consists of a proton plus a negative meson. Though the
total charge remains zero, a charge density can result,
extending a distance of about $1/m_\pi$ from the center of mass
of the neutron. Let us say the charge density is $\rho(\underline{r}_1)$, $\underline{r}_1$
being measured from the center of mass. If the
electrostatic potential in an atom at point $\underline{r}'$ is $\emptyset(\underline{r}')$,
the neutron with center of mass at $\underline{r}$ feels a potential

$$V(\underline{r})=\int\rho(\underline{r}'-\underline{r})\emptyset(\underline{r}')d\underline{r}', \tag{6.29}$$

and the scattering amplitude of the atom will be (as in
(6.25))

$$f_{\underline{q}}=-a-(M/2\pi)V_{\underline{q}}. \tag{6.30}$$

The folding theorem tells us that[7]

$$V_{\underline{q}}=\rho_{\underline{q}}\emptyset_{-\underline{q}}. \tag{6.31}$$

Since

$$\Delta\emptyset=-4\pi\rho',$$

with ρ' the charge density of the atom, we have

$$\emptyset_{\underline{q}}=4\pi\rho'_{\underline{q}}/q^2,$$

---

[a] W.W. Havens, J.Rainwater and I.I. Rabi, Phys. Rev. $\underline{72}$,
634 (1947); E. Fermi and L. Marshall, Phys. Rev. $\underline{72}$, 1139
(1947).

[7] To prove, write
$$V(\underline{r})=1/(2\pi)^6\int\rho_{\underline{q}}e^{i\underline{q}\cdot(\underline{r}'-\underline{r})}\emptyset_{\underline{q}'}e^{i\underline{q}'\cdot\underline{r}'}d\underline{q}d\underline{q}'d\underline{r}'.$$
The $\underline{r}'$ integration gives $(2\pi)^3\delta(\underline{q}+\underline{q}')$, whence
$$V(\underline{r})=1/(2\pi)^3\int\rho_{\underline{q}}\emptyset_{-\underline{q}}e^{-i\underline{q}\cdot\underline{r}}d\underline{q}.$$

and (6.31) can be written

$$V_{\underline{q}} = 4\pi \rho_{\underline{q}} \rho'_{-\underline{q}} / q^2. \tag{6.32}$$

Now

$$\rho_{\underline{q}} = \int \rho(\underline{r}) e^{-i\underline{q} \cdot \underline{r}} d\underline{r},$$

and since (for thermal neutron scattering) the magnitude of $\underline{q}$ times the radius of the neutron charge distribution is much smaller than one, the exponential factor can be expanded in a rapidly converging power series,

$$\rho_{\underline{q}} = \int \rho(\underline{r}) [1 - i(\underline{q} \cdot \underline{r}) - (1/2)(\underline{q} \cdot \underline{r})^2] d\underline{r}.$$

The first term vanishes because the total charge is zero, the second because the neutron can have no electric dipole moment if parity is conserved. Furthermore, a spin 1/2 particle can have no quadrupole or higher moments, so $\rho(\underline{r})$ must be spherically symmetric. We then have

$$\rho_{\underline{q}} = -(q^2/6) \int \rho(\underline{r}) r^2 d\underline{r} = (e/6) \langle r^2 \rangle q^2, \tag{6.33}$$

a relationship defining the mean-square charge radius, $\langle r^2 \rangle$ (since we expect $\rho(r)$ to be negative for $r > 0$, we remove a factor $-e$). Equation (6.32) can then be written

$$V_{\underline{q}} = -(2\pi/3) e \langle r^2 \rangle \rho'_{-\underline{q}}. \tag{6.34}$$

The contribution of the atom's nuclear charge to $\rho'_{-\underline{q}}$ is, since for the values of q in question q times the nuclear radius is small, simply Ze. This gives a constant term in (6.30) which, in principle, is already included in the nuclear scattering length a. So the charge density in (6.34) should be taken to be that of the atomic electrons. If we write this in terms of a form factor,

$$\rho'_{-\underline{q}} = -ZeF(\underline{q}),$$

with F(0)=1, we have

$$V_{\underline{q}} = -(2\pi/3) Ze^2 \langle r^2 \rangle F(\underline{q}) \tag{6.35}$$

and

$$f_{\underline{q}} = -[a - (M/3) Ze^2 \langle r^2 \rangle F(\underline{q})]. \tag{6.36}$$

The value of $\langle r^2 \rangle$ defined by (6.33) is not a
completely geometrically determined quantity, for example
in the meson theory it evidently is proportional to the
probability of the neutron being dissociated into proton
and π^-. To estimate the expected effect, Fermi assumed
$$\langle r^2 \rangle = p_\pi / m_\pi^2,$$
where p_π is the probability of dissociation. He determined
p_π from the neutron magnetic moment, assuming that in the
dissociated configuration the π^- is in a p state. Using the
vector model, a calculation similar to that of the deuteron
magnetic moment in Sec.6III gives (expressed in nuclear
magnetons, so $\mu_\pi = -M/m_\pi$ in the p state)
$$\mu_N = -(p_\pi/3)(\mu_p + 2M/m_\pi),$$
From this relationship we find $p_\pi = 0.35$. Fermi, however,
gave $p_\pi = 0.2$, which would result from taking the dissociated
state of the neutron to be $m_s = -1/2$, $m_l = 1$, rather than the
state of angular momentum 1/2, i.e. the state $m_s = 1/2$, $m_l = 0$
was ignored. Equation (6.35) becomes

$$f_q = -[a - (\alpha Z p_\pi M/3m_\pi^2)F(q)], \qquad (6.37)$$

where $\alpha = e^2$ is the fine structure constant. It is evident
that it is desirable to use a high Z element to take
advantage of the coherent scattering of many electrons.

Fermi and Marshall used Xe ($Z=54$), for which $a=5.9$ f,
so (6.37) becomes (using Fermi's $p_\pi = 0.2$)
$$f_q = -(5.9 - 0.25F(q)).$$
The form factor $F(q)$ had been calculated by Bethe, and
Fermi and Marshall attempted to detect the predicted
angular dependence of the neutron scattering. They failed
to detect an effect. The experimenters used a convention of
expressing the neutron-electron interaction in terms of the
depth, V, of an attractive square well potential of range
equal to the classical electron radius, $r_0 = e^2/m$, which
would give the same forward scattering amplitude. For a

square well the $q=0$ Fourier component is
$$V_0 = \int V(r)\,d\underline{r} = -(4\pi r_0^3/3)V.$$
Equating this to (6.35) with $q=0$, $Z=1$ gives
$$V = (1/2)(\langle r^2 \rangle / r_0^2)m. \qquad (6.38)$$
If we put $\langle r^2 \rangle = p_\pi / m_\pi^2$ we obtain, using Fermi's value of p_π,
$$V = (1/2)p_\pi(1/m_\pi^2 r_0^2)m = (1/2)\cdot 0.2 \cdot (1.4/2.8)^2 \cdot .51 \text{ Mev} = 13 \text{ Kev}$$
as the "meson theory" value. In these terms, Fermi and
Marshall's experimental result was $V=-300\pm5000$ Kev.

Havens, Rabi and Rainwater used a different
experimental method. They measured the total cross section
in liquid Pb as a function of neutron energy. If we write
$$f_q = -[a - Za_{el}F(q)],$$
the total cross section is
$$\sigma_{tot} = 4\pi[a^2 - 2Za_{el}\langle F(q) \rangle] + \sigma_{cap}.$$
The average form factor, $\langle F(q) \rangle$, can be calculated as a
function of energy from the $F(q)$ of the Fermi-Thomas atomic
model. The measured cross sections have to be corrected for
the energy dependence of the capture cross section, for the
thermal motion of the target atoms (Sec.3II), for the
density correlation of atoms in the liquid (Sec.3IV) and,
possibly, by addition of a spin-flip cross section if the
nuclear spin is not zero (Sec.3III). The conclusion was
that $V \approx 2.5$ Kev. A later experiment[*], using liquid Bi, found
a 3% increase in cross section when the neutron wavelength
was varied from $\lambda=1.33$ A to $\lambda=0.33$ A. About 40% of the
increase was accounted for by the corrections mentioned
above. The remainder led to the result
$$V = 5300 \pm 1000 \text{ ev}.$$

---

[*] W.W. Havens Jr., J.Rainwater and I.I. Rabi, Phys. Rev.
82, 845 (1951)

Just after the Bi result was published Foldy[*] pointed
out that if the neutron is treated as a Dirac particle
there is an effect which leads to a neutron—electron
interaction, namely, a spin—orbit interaction. One would
expect the interaction of the neutron magnetic moment with
the magnetic field to be

$$V=-\mu_N(e/2M)\underline{\sigma}\cdot\underline{H}',$$

where $\underline{H}'$ is the field in the coordinate system in which the
neutron is at rest, or, in terms of fields in the
laboratory system

$$V=-\mu_N(e/2M)\underline{\sigma}\cdot(\underline{H}-\underline{v}\times\underline{E}). \qquad (6.39)$$

We are concerned with the second term, which represents the
spin—orbit interaction. Since the observation of the
neutron—electron interaction depends on the interference
term with the nuclear scattering, which is linear in V, one
would at first sight expect no effect for an unpolarized
neutron beam. However, one should recall that the Dirac
theory gives, in the case of the atomic fine structure, a
non-zero spin—orbit interaction for s states. In the Dirac
theory it may happen that $\underline{v}$ is replaced by the operator $\underline{\alpha}$,
with unexpected results. Since $\underline{\alpha}$ and $\underline{\sigma}$ do not commute it is
not immediately obvious how the interaction should be
written in the Dirac theory. The simplest way to find the
correct form is to write the interaction in a Lorentz
invariant form. If ψ is an arbitrary solution of the Dirac
equation, the expectation value of V can be written

$$\langle V\rangle=\mu_N(e/2M)\int\psi^*\gamma_4(1/2)\gamma_\mu\gamma_\nu F_{\mu\nu}\psi d\underline{r},$$

the so-called Pauli interaction. Here, in terms of two by
two submatrices,

$$\gamma_4=\left|\begin{array}{c|c} 1 & 0 \\ \hline 0 & -1 \end{array}\right|, \qquad \gamma_i=\left|\begin{array}{c|c} 0 & -i\sigma_i \\ \hline i\sigma_i & 0 \end{array}\right|, \qquad i=1,2,3,$$

---

[*] L.L. Foldy, Phys. Rev. **83**, 688 (1951)

$F_{ij}=H_k$, $i,j=1,2,3$, ijk in cyclic order, $F_{4i}=iE_i$. Noting that $\gamma_i\gamma_j=i\sigma_k$, we can rewrite this as

$$\langle V\rangle=-\mu_N(e/2M)\int\psi^*\gamma_4(\sigma\cdot H+\gamma_4\gamma\cdot E)\psi d\underline{r}.$$

For $H=0$,

$$\langle V\rangle=-\mu_N(e/2M)\int\psi^*\gamma\cdot E\psi d\underline{r}.$$

If we write the four-component Dirac ψ in terms of two two-component spinors,

$$\psi=\left|\frac{\phi_1}{\phi_2}\right| ,$$

we have

$$\langle V\rangle=i\mu_N(e/2M)\int[\phi_1^*\sigma\cdot E\phi_2-\phi_2^*\sigma\cdot E\phi_1]d\underline{r}.$$

In the non-relativistic limit the Dirac equation gives

$$\phi_2=(1/2M)\sigma\cdot\underline{p}\phi_1,$$

with $\underline{p}=-i\mathrm{grad}$, so

$$\langle V\rangle=i\mu_N(e/4M^2)\int\phi_1^*[(\sigma\cdot E)(\sigma\cdot\underline{p})-(\sigma\cdot\underline{p})(\sigma\cdot E)]\phi_1 d\underline{r}.$$

From the relation $(\sigma\cdot A)(\sigma\cdot B)=A\cdot B+i\sigma\cdot(A\times B)$ we obtain for the square bracket

$$[\quad]=E\cdot\underline{p}-\underline{p}\cdot E+i\sigma\cdot(E\times\underline{p}-\underline{p}\times E).$$

The first two terms are the commutators

$$\Sigma_i[E_i,p_i]=i\mathrm{div}E.$$

The last two equal

$$-2i\sigma\cdot(\underline{p}\times E),$$

since the difference between $E\times\underline{p}$ and $\underline{p}\times E$ equals $-i\mathrm{curl}E=0$. Since $\underline{p}/M=\underline{v}$, this last term gives a result the same as the second term in (6.39). There remains the div E term, which gives

$$\langle V\rangle=-\mu_N(e/4M^2)\int\phi_1^*\mathrm{div}E\phi_1 d\underline{r},$$

or, since $\mathrm{div}E=4\pi\rho'$, an interaction potential

$$V=-\mu_N(\pi e/M^2)\rho'.$$

Thus

$$V_{\underline{q}}=-\mu_N(\pi e/M^2)\rho_{\underline{q}}'=\mu_N(\pi e^2 Z/M^2)F(\underline{q}), \qquad (6.40)$$

and (6.30) becomes

$$f_{\underline{q}}=-[a-Za_{el}F(\underline{q})]$$

with

$$a_{el}=-\mu_N(e^2/2M)=1.467\times10^{-3} f.$$

For example, for Bi with $Z=83$ and a coherent scattering length $a=8.625$ f,

$$f_g = -(8.625-0.1218F(g),$$

a correction of 1.4% to the forward amplitude.

The equivalent square well potential depth is

$$V=-(3/4)\mu_N(e^2/M^2r_0^3)=-(3/4)\mu_N(1/M^2r_0^2)m=4073 \text{ ev},$$

not too different from the result of Havens, Rabi and Rainwater. An improved version of the same experiment[10] gave

$$V=4165\pm265 \text{ ev}.$$

A different experimental method was used by Hughes, Harvey and Goldberg[11], who measured the critical angle for total internal reflection at a Bi-liquid oxygen interface. The critical angle is proportional to the change in Nf_0 across the interface. The coherent nuclear scattering lengths, a, were determined by scattering measurements for 8 ev neutrons, for which $\langle F(g) \rangle \approx 0$. The Bi-liquid oxygen combination was chosen because Na for liquid oxygen differs by only 2% from Na for Bi, so that in the difference of Nf_0's the electron terms become of comparable magnitude to the nuclear terms. The result was

$$V=3860\pm370 \text{ ev}.$$

The weighted mean of the last two results gives

$$V=4062\pm215 \text{ ev},$$

agreeing perfectly with the spin-orbit prediction.

It is not accidental that (6.40) and (6.34) are of the same form. If we rewrite (6.39) as

$$V=-\mu_N(e/2M)[\underline{\sigma}\cdot\underline{H}+(\underline{v}\times\underline{\sigma})\cdot\underline{E}]$$

[10] E. Melkonian, B.M. Rustad and W.W. Havens Jr., Bull. Am. Phys. Soc. 1, 62 (1956)
[11] D.J. Hughes, J.A. Harvey and M.D. Goldberg, Phys. Rev. 90, 497 (1956)

it reminds us that another description of the spin-orbit
interaction is that a moving magnetic dipole distribution
$\underline{M}(\underline{r})$ gives rise to an electric dipole density $\underline{P}=\underline{v}\times\underline{M}(\underline{r})$. In
the Dirac theory $\underline{P}=\mu_N (e/2M)\psi^* \underline{\gamma}\psi$. This is equivalent to a
charge density $\rho=\text{div}\underline{P}$. The experiments show the presence of
no charge density other than that associated with the
magnetic moment.

CHAPTER 7

P-P AND N-N INTERACTIONS

Heisenberg's 1932 nuclear model considered only N-P and N-N forces. Direct evidence of P-P forces required P-P scattering experiments (there is no bound state), and, because of the Coulomb repulsion between the protons, these awaited the development of particle accelerators which could produce proton beams in the 1 Mev range. In classical mechanics the P-P forces could not be detected until the distance of closest approach in the collision became smaller than the range, R, of the nuclear force, i.e. until the center of mass energy became greater than the barrier height,

$$E_{cm} > e^2/R.$$

If we set R equal to the N-P singlet effective range (since the P-P s state must be ^1S), $R = r_s = 2.6$ f, we find

$$E_{cm} > 0.55 \text{ Mev},$$

or $E_{lab} > 1.1$ Mev. Because of the quantum mechanical barrier penetration effect the needed energy would actually be

somewhat lower. The first experiment which detected the P-P
forces was carried out by M.G. White[1], who observed the
scattering in a cloud chamber of 0.7 Mev protons from the
Berkekey 37" cyclotron. At 90° (center of mass system) he
observed ten times the cross section expected for Coulomb
scattering alone. White's experiment just anteceded the
first measurement of the low energy N-P cross section by
Dunning, Pegram, Fink and Mitchell, which was published in
the following volume of the Physical Review.

---

[1] M.G. White, Phys. Rev. 47, 573 (1935).

SEC.7I COULOMB SCATTERING

The analysis of P-P scattering is more complicated than that of N-P because scattering by the Coulomb field must be taken into account. If we attempt to solve the Coulomb scattering problem in the obvious way, separating the wave equation in spherical coordinates, we are faced with performing the sum

$$f(\theta)=(1/2ip)\Sigma_l(e^{2i\delta_l}-1)P_l(\cos\theta);$$

since the range of the force is infinite this cannot be expected to be a rapidly converging series. This difficulty could be circumvented by using the Born approximation (provided it is valid). Born approximation gives

$$f(\theta)=-(\mu/2\pi)V_q,$$

with μ the reduced mass and q the momentum transfer. For scattering of a particle of charge Z_1e by another of charge Z_2e the Coulomb potential is

$$V=Z_1Z_2e^2/r$$

and the Fourier component is

$$V_q=4\pi Z_1Z_2e^2/q^2.$$

This gives

$$f(\theta)=-2\mu Z_1Z_2e^2/q^2.$$

Since $q=p'-p$, we have $q^2=2p^2(1-\cos\theta)=4p^2\sin^2(\theta/2)$, and

$$f(\theta)=-Z_1Z_2e^2/2\mu v^2\sin^2(\theta/2),$$

$$d\sigma=\frac{(Z_1Z_2e^2)^2}{(2\mu v^2)^2\sin^4(\theta/2)}d\Omega,$$

which is just the Rutherford formula.

A difficulty appears, however, if we investigate the validity of the Born approximation. If we write the Schrodinger equation

$$\Delta\psi+p^2\psi=2\mu V\psi$$

we see that ψ satisfies the integral equation

$$\psi(r)=e^{ip\cdot r}-(2\mu/4\pi)\int(e^{ip|r-r'|}/|r-r'|)V(r')\psi(r')dr'.$$

The Born approximation is obtained by approximating $\psi(r')$

by $e^{i\underline{p}\cdot\underline{r}'}$,

$$\psi(\underline{r})=e^{i\underline{p}\cdot\underline{r}}-(\mu/2\pi)\int(e^{ip|\underline{r}-\underline{r}'|}/|\underline{r}-\underline{r}'|)V(\underline{r}')e^{i\underline{p}\cdot\underline{r}'}d\underline{r}',\quad(7.01)$$

and taking the asymptotic form for large $\underline{r}$. Its validity thus depends on on the second term remaining small compared to the first in the region where V is large and the scattering takes place. The simplest point to check is $\underline{r}=0$, where (7.01) becomes

$$\psi(0)=1-(\mu/2\pi)\int(e^{ipr'}/r')V(\underline{r}')e^{i\underline{p}\cdot\underline{r}'}d\underline{r}'.$$

For the Born approximation to be valid the magnitude of the second term must be small compared to one. For a spherically symmetric potential the condition is

$$|(2\mu/p)\int_0^\infty e^{ipr'}\sin pr' V(r')dr'|<<1,$$

or, writing the sine in terms of exponentials,

$$|(\mu/p)\int_0^\infty(e^{2ipr'}-1)V(r')dr'|<<1.$$

Unless V falls off faster than $1/r$ the integral doesn't even converge. If we take a shielded Coulomb potential, with a spatial dependence $e^{-\lambda r}/r$, the quantity whose magnitude we require is

$$(Z_1Z_2e^2\mu/p)\int_0^\infty(e^{-(\lambda-2ip)r'}-e^{-\lambda r'})dr'/r'.$$

The integral can be evaluated by writing it as

$$\lim_{\epsilon\to 0}[\int_\epsilon^\infty e^{-(\lambda-2ip)r'}dr'/r'-\int_\epsilon^\infty e^{-\lambda r'}dr'/r']$$

$$=\lim_{\epsilon\to 0}[\int_{(\lambda-2ip)\epsilon}^\infty e^{-x}dx/x-\int_{\lambda\epsilon}^\infty e^{-x}dx/x]$$

$$=\lim_{\epsilon\to 0}\int_{(\lambda-2ip)\epsilon}^{\lambda\epsilon}e^{-x}dx/x=\int_{(\lambda-2ip)\epsilon}^{\lambda\epsilon}dx/x=\ln[\lambda/(\lambda-2ip)].$$

For large p/λ the condition for the validity of the Born approximation becomes

$$|Z_1Z_2e^2/v|\ln(p/\lambda)<<1.$$

For a Coulomb field $(\lambda=0)$ the condition is , of course, violated. Even for finite λ, it is violated unless

$$|Z_1Z_2e^2/v|<<1.$$

Let us write

$$\gamma=Z_1Z_2e^2/v.$$

In c.g.s. units $\gamma = 2\pi Z_1 Z_2 e^2 / hv$; thus in the classical limit, $h \to 0$, γ is large. For positive γ, the classical distance of closest approach is $r_0 = Z_1 Z_2 e^2 2\mu / p^2$, and

$$\gamma = (1/2) p r_0.$$

Thus γ large means the distance of closest approach is large compared to the wavelength, so a wave-packet can be constructed which follows the classical trajectory. If γ is negative the system has hydrogenic bound states; γ is the ratio of the wavelength, $1/p$, to the radius of the 1s state, and $\gamma = 1$ corresponds to E equal in magnitude to the 1s binding energy.

The Rutherford formula is the correct answer, also in quantum mechanics. The Born approximation does give the scattering amplitude correctly to lowest order in γ; the failure of the test indicates a more deep-lying difficulty: the usual asymptotic form of the wave function for large r,

$$\psi \approx e^{ipz} + f(\theta) e^{ipr} / r, \tag{7.02}$$

is correct only for a $V(r)$ falling off faster than $1/r$. This can be seen by solving the radial wave function in the W.K.B. approximation, which is certainly correct for large r. The W.K.B. radial function for angular momentum l is

$$u_l(r) = \sin\{\int_{r_0}^r [2\mu(E-V) - (l+1/2)^2 / r^2]^{1/2} dr + \pi/4\},$$

where r_0 is the turning point. For a Coulomb potential it becomes

$$u_l(r) = \sin\{\int_{r_0}^r [p^2 - (2\mu Z_1 Z_2 e^2 / r) - (l+1/2)^2 / r^2]^{1/2} dr + \pi/4\}. \tag{7.03}$$

For large r we can expand the square root in powers of $1/r$; to order $1/r$ the result is

$$\int (p - \gamma/r) dr = pr - \gamma \ln r + const.$$

Even without expanding the integral is elementary, and we can readily find the asymptotic form of u_l,

$$u_l \approx \sin[pr - \gamma \ln(2pr) - (\pi/2)l + \eta_l],$$

with

$$\eta_1 = \gamma\{\ln[(1+1/2)^2+\gamma^2]^{1/2}-1\}+(1+1/2)\sin^{-1}\{\gamma/[(1+1/2)^2+\gamma^2]^{1/2}\}.$$

(7.04)

If we write this in terms of exponentials we see that the outgoing wave is not of the form (7.02), but rather
$$f(\theta)e^{i(pr-\gamma\ln(2pr))}/r.$$

To investigate the plane wave term in (7.02) let us look at the three dimensional W.K.B. approximation. An easy way to formulate this is to write the variation principle

$$\delta\int[(1/2\mu)\text{grad}\psi^*\cdot\text{grad}\psi+\psi^*V\psi-E\psi^*\psi]d\underline{r}=0.$$

(7.05)

Independent variations of ψ^* and ψ give the Schrodinger equation and its complex conjugate. If we take the form
$$\psi=A(\underline{r})e^{iS(\underline{r})},$$

with A and S real, (7.05) becomes

$$\delta\int\{(1/2\mu)(\text{grad}A)^2+A^2[(1/2\mu)(\text{grad}S)^2+V-E]\}d\underline{r}=0.$$

(7.06)

The W.K.B. approximation is obtained by supposing that A is slowly varying, and neglecting the $(\text{grad}A)^2$ term. The result of varying A then gives
$$(1/2\mu)(\text{grad}S)^2+V-E=0,$$
which is just the classical Hamilton-Jacobi equation. The momentum at any point is

$$\underline{p}=\text{grad}S;$$

thus the trajectories are normal to the surfaces S=const. Varying S gives the equation which determines A^2,
$$\text{div}(A^2\text{grad}S)=0, \quad \text{i.e.} \quad \text{div}(A^2\underline{p})=0,$$

which is just the current conservation law.

The condition for the validity of the W.K.B. approximation is
$$(\text{grad}A)^2\ll A^2(\text{grad}S)^2=A^2p^2,$$

which says that A varies by only a small fraction of itself in a wavelength.

For the Coulomb field problem the Hamilton-Jacobi equation is known to separate not only in spherical coordinates, but in parabolic coordinates as well. This set

of orthogonal coordinates is defined by the relations
$$x=(\mathcal{J}'\eta')^{1/2}\cos\varnothing, \quad y=(\mathcal{J}'\eta')^{1/2}\sin\varnothing, \quad z=(1/2)(\mathcal{J}'-\eta').$$
From these it follows that
$$r=(1/2)(\mathcal{J}'+\eta'), \quad \rho^2=x^2+y^2=\mathcal{J}'\eta',$$
$$\mathcal{J}'=r+z=r(1+\cos\varnothing), \quad \eta'=r-z=r(1-\cos\varnothing)$$

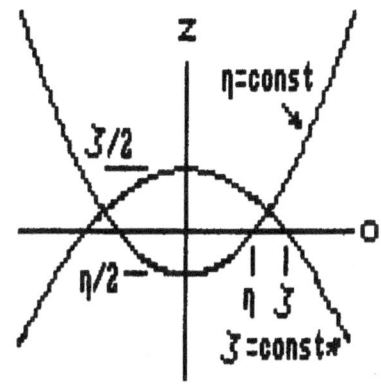

The surfaces $\mathcal{J}'$ or η' constant are paraboloids of revolution. We can write
$$z=(1/2)(\mathcal{J}'-\rho^2/\mathcal{J}'),$$
so the surfaces of constant $\mathcal{J}'$ are the parabolas shown in Fig7.1. Similarly
$$z=(1/2)(\rho^2/\eta'-\eta'),$$
which gives the parabolas shown for η'=const.

Fig.7.1

For any orthogonal coordinates, q_i, the metric is
$$ds^2=dx^2+dy^2+dz^2=\Sigma_i h_i^2 dq_i^2,$$
where
$$h_i^2=(\partial x/\partial q_i)^2+(\partial y/\partial q_i)^2+(\partial z/\partial q_i)^2,$$
and the components of gradS are
$$(1/h_i)\partial S/\partial q_i,$$
so the Hamilton-Jacobi equation becomes
$$\Sigma_i(1/h_i^2)(\partial S/\partial q_i)^2+2\mu(V-E)=0.$$
For parabolic coordinates one easily finds
$$h_{\mathcal{J}'}^2=(1/4)(\mathcal{J}'+\eta')/\mathcal{J}', \quad h_{\eta'}^2=(1/4)(\mathcal{J}'+\eta')/\eta', \quad h_\varnothing^2=\mathcal{J}'\eta'.$$
For
$$V=Z_1 Z_2 e^2/r=2Z_1 Z_2 e^2/(\mathcal{J}'+\eta')$$
we can take S independent of $\varnothing$ (the classical orbit is then in a plane $\varnothing$=const), and the Hamilton-Jacobi equation is
$$\mathcal{J}'(\partial S/\partial\mathcal{J}')^2+\eta'(\partial S/\partial\eta')^2+Z_1 Z_2 e^2\mu-(1/4)p^2(\mathcal{J}'+\eta')=0.$$
In this equation $p^2=2\mu E$.

If we rescale this equation by introducing the dimensionless variables

$$\mathfrak{s}=p\,\mathfrak{s}', \quad \eta=p\eta',$$

the equation becomes

$$\mathfrak{s}(\partial S/\partial \mathfrak{s})^2+\eta(\partial S/\partial \eta)^2+\gamma-(1/4)(\mathfrak{s}+\eta)=0.$$

It is evident that the variables separate; if we write

$$S=S_1(\mathfrak{s})+S_2(\eta)$$

we have

$$\mathfrak{s}(dS_1/d\mathfrak{s})^2-(1/4)\mathfrak{s}+\omega=0, \qquad \eta(dS_2/d\eta)^2-(1/4)\eta+\nu=0,$$

with

$$\omega+\nu=\gamma.$$

Thus

$$p_{\mathfrak{s}}=dS_1/d\mathfrak{s}=\pm(1/2)[(\mathfrak{s}-4\omega)/\mathfrak{s}]^{1/2},$$

$$p_{\eta}=dS_2/d\eta=\pm(1/2)[(\eta-4\nu)/\eta]^{1/2}. \qquad (7.07)$$

We have not yet applied the condition that the initial velocity is in z direction. This means that as $z\to-\infty$ the trajectory approaches a constant ρ. Since $z=(1/2)(\mathfrak{s}-\eta)/p$ and $\rho=\sqrt{(\mathfrak{s}\eta)}/p$, $z\to-\infty$ requires $\eta\to\infty$, and ρ constant then requires $\mathfrak{s}\to0$. Since it is evident from (7.07) that $\mathfrak{s}\geq4\omega$, $\mathfrak{s}\to0$ means $\omega\leq0$. If $p_{\mathfrak{s}}$ is to remain finite as $\mathfrak{s}\to0$ we must have $\omega=0$.[2] Also, we must take the + sign in (7.07) for $p_{\mathfrak{s}}$, since $\mathfrak{s}$ can only increase from 0. On the other hand, we must take the − sign for p_{η}, since $-d\eta/dt$ appears in the expression for dz/dt. Thus

$$p_{\eta}=dS_2/d\eta=-(1/2)[(\eta-4\gamma)/\eta]^{1/2}.$$

As we proceed along the trajectory η decreases until it reaches a turning point, either $\eta=4\gamma$ if γ is positive or $\eta=0$ if γ is negative. Beyond this the sign of p_{η} is reversed, and η increases again. For definiteness, suppose

---

[2] It is dynamically allowed for $p_{\mathfrak{s}}$ to become infinite like $1/\sqrt{\mathfrak{s}}$, since the associated kinetic energy is

$$(1/2\mu h_{\mathfrak{s}}^2)p_{\mathfrak{s}}^2=(2/\mu)[\mathfrak{s}/(\mathfrak{s}+\eta)]p_{\mathfrak{s}}^2.$$

But then the point $\mathfrak{s}=0$ is a turning point of the trajectory, where the square root in (7.07) changes sign. At this point $\rho=0$ and $d\rho/dt$ changes sign. It simply means that the trajectory crosses the negative z axis, so the initial velocity is not parallel to z.

$\gamma>0$. Then

$$S_2 = -\{\int_{4\gamma}^{\eta}(1/2)[(\eta-4\gamma)/\eta]^{1/2}d\eta+\pi/4\},$$

if we use the usual W.K.B. recipe for the phase at the turning point, as in (7.03). The integral gives

$$S_2 = -(1/2)[\eta(\eta-4\gamma)]^{1/2}+2\gamma\ln[\eta^{1/2}+(\eta-4\gamma)^{1/2}]-\gamma\ln4\gamma-\pi/4.$$

For large η

$$S_2 \approx -(1/2)\eta+\gamma\ln\eta-\eta_0-\gamma^2/\eta,$$

where

$$\eta_0 = \gamma(\ln\gamma-1)+\pi/4. \qquad (7.08)$$

We also have

$$dS_1/d\int=1/2, \quad S_1=(1/2)\int,$$

so

$$S \approx (1/2)(\int-\eta)+\gamma\ln\eta-\eta_0-\gamma^2/\eta.$$

To obtain the equation for A^2 in curvilinear coordinates, write (7.06) (with the $(\text{grad}A)^2$ term neglected) as

$$\delta\int A^2[\frac{1}{2\mu}\sum_i\frac{1}{h_i^2}(\frac{\partial S}{\partial q_i})^2+V-E]hdq_i ,$$

where $h=h_1h_2h_3$. Varying S gives the equation

$$\sum_i\frac{\partial}{\partial q_i}(\frac{h}{h_i^2}\frac{\partial S}{\partial q_i}A^2)=0,$$

in our case

$$\frac{\partial}{\partial\int}(\int\frac{\partial S}{\partial\int}A^2)+\frac{\partial}{\partial\eta}(\eta\frac{\partial S}{\partial\eta}A^2)=0$$

Since $\partial S/\partial\int=1/2$, we can find a solution $A=A(\eta)$, with

$$\frac{\partial}{\partial\eta}(\eta\frac{\partial S}{\partial\eta}A^2)+(1/2)A^2=0.$$

For large η, $\partial S/\partial\eta=(1/2)$ and the equation is

$$(d/d\eta)(\eta A^2)-A^2=0, \qquad (7.09)$$

with a solution $A^2=1$.

Thus for large η the W.K.B. approximation takes the form

$$\psi \approx e^{i[(1/2)(\int-\eta)+\gamma\ln\eta-\eta_0-\gamma^2/\eta]},$$

or, since

$$\eta=pr(1-\cos\theta)=2pr\sin^2(\theta/2),$$

$$\psi \approx e^{i[pz+\gamma\ln(2pr\sin^2(\theta/2))-\eta_0-\gamma^2/2pr\sin^2(\theta/2)]}.$$

This shows how the plane wave term of (7.02) must be modified. The trouble is that for the infinite range Coulomb potential the surfaces of constant S, normal to the trajectories, do not approach planes no matter how far from the scattering center.

The argument can be carried further: beyond the turning point the signs of p_η and S_2 are reversed. Thus for large η after the scattering

$$S \approx (1/2)(\mathcal{J}+\eta) - \gamma \ln \eta + \eta_0 + \gamma^2/\eta.$$

Instead of (7.09) we have

$$(d/d\eta)(\eta A^2) + A^2 = 0,$$

which has the solution

$$A^2 = c^2/\eta^2.$$

We may infer that the asymptotic form of ψ is the sum of the W.K.B. solutions,

$$\psi \approx e^{i[(1/2)(\mathcal{J}-\eta)+\gamma \ln \eta - \eta_0 - \gamma^2/\eta]} + (c/\eta)e^{i[(1/2)(\mathcal{J}+\eta)-\gamma \ln \eta + \eta_0]}.$$

In the classical limit, $\gamma \gg 1$, one should obtain the Rutherford cross section. This is achieved if $c = -\gamma$ (we take the minus sign because one expects a positive potential to produce a negative scattering amplitude). Then we can write

$$\psi \approx e^{-i\eta_0}\{ e^{i[pz+\gamma \ln(2prsin^2(\theta/2)) - \gamma^2/2prsin^2(\theta/2)]} + f(\theta)e^{i[pr-\gamma \ln(2pr)]}/r\},$$

with

$$f(\theta) = -(\gamma/2psin^2(\theta/2))e^{-i\gamma \ln(sin^2(\theta/2))+2i\eta_0}.$$

This differs from the Born approximation result by a phase factor which, no matter how small γ, becomes appreciable for small enough θ.

Consideration of the W.K.B. approximation is useful for purposes of orientation. Actually, the Schrodinger equation can be solved exactly in closed form with the use of parabolic coordinates.

The easiest way to express the the Schrodinger equation in terms of orthogonal curvilinear coordinates (i.e. obtain the proper expression for the Laplacian operator) is to write the variation principle, (7.05), as

$$\delta \int [\frac{1}{2\mu} \sum_i \frac{1}{h_i^2} \frac{\partial \psi^*}{\partial q_i} \frac{\partial \psi}{\partial q_i} + \psi^* (V-E)\psi]hd\underline{q}.$$

Taking the variation with respect to ψ^* gives the equation

$$[\frac{1}{2\mu h} \sum_i \frac{\partial}{\partial q_i} \frac{h}{h_i^2} \frac{\partial}{\partial q_i} + (E-V)]\psi = 0.$$

Thus

$$\Delta = \frac{1}{h} \sum_i \frac{\partial}{\partial q_i} \frac{h}{h_i^2} \frac{\partial}{\partial q_i},$$

which, for parabolic coordinates, is

$$\Delta = \frac{4}{\xi + \eta} (\frac{\partial}{\partial \xi} \xi \frac{\partial}{\partial \xi} + \frac{\partial}{\partial \eta} \eta \frac{\partial}{\partial \eta}) + \frac{1}{\xi \eta} \frac{\partial^2}{\partial \theta^2}.$$

For a Coulomb potential and a ψ independent of $\emptyset$, the wave equation in scaled variables is

$$[\frac{\partial}{\partial \xi} \xi \frac{\partial}{\partial \xi} + \frac{\partial}{\partial \eta} \eta \frac{\partial}{\partial \eta} + (1/4)(\xi + \eta) - \gamma]\psi = 0.$$

The variables separate; if we write

$$\psi = u(\xi) v(\eta)$$

we obtain

$$[\frac{\partial}{\partial \xi} \xi \frac{\partial}{\partial \xi} + (1/4)\xi - \omega]u = 0,$$

$$[\frac{\partial}{\partial \eta} \eta \frac{\partial}{\partial \eta} + (1/4)\eta - \nu]v = 0, \qquad (7.10)$$

with $\omega + \nu = \gamma$.

These equations have a regular singular point at the origin. The roots of the indicial equation are both zero, so there is a regular solution which approaches a constant at the origin and an irregular solution which behaves like $\ln \xi$ (or $\ln \eta$).

To solve the differential equations we shall use a method which is often useful for equations with polynomial coefficients. We write

$$u(\xi) = \int_{t_1}^{t_2} e^{-\xi t} w(t)dt,$$

where the integral is to be taken along some path in the complex t plane. If we write (7.10) as

$$Lu(\int)=0,$$

we find

$$Lu(\int)=\int_{t_1}^{t_2} e^{-\int t} w(t)[\int(t^2+1/4)-(t+\mu)]dt,$$

$$=\int_{t_1}^{t_2}\{(-de^{-\int t}/dt)w(t)(t^2+1/4)-e^{-\int t}w(t)(t+\mu)\}dt,$$

$$=-e^{-\int t}w(t)(t+1/4)\Big|_{t_2}^{t_1}$$

$$+\int_{t_2}^{t_1} e^{-\int t}\{(d/dt)[w(t)(t^2+1/4)]-w(t)(t+u)\}dt=0.$$

The differential equation will be satisfied if t_1 and t_2 are chosen so the first term vanishes and w(t) is chosen so that the integrand of the second term vanishes. If we write

$$\emptyset(t)=w(t)(t^2+1/4)$$

the the latter condition can be written

$$d\emptyset/dt=[(t+w)/(t^2+1/4)]\emptyset,$$

or

$$\frac{d\emptyset}{\emptyset}=\frac{tdt}{t^2+1/4}+i\omega(\frac{1}{t+i/2}-\frac{1}{t-i/2})dt,$$

whence

$$\ln\emptyset=(1/2)\ln(t^2+1/4)+i\omega\ln[(t+i/2)/(t-i/2]+const,$$

$$\emptyset=c(t+i/2)^{(1/2)+i\omega}(t-i/2)^{(1/2)-i\omega},$$

$$w=c(t+i/2)^{-(1/2)+i\omega}(t-i/2)^{-(1/2)-i\omega}.$$

If we choose $c=1/2\pi i$, we have a solution

$$u(\int)=(1/2\pi i)\int_{t_1}^{t_2} e^{-\int t}(t+i/2)^{-(1/2)+i\omega}(t-i/2)^{-(1/2)-i\omega}dt,$$

$$(7.11)$$

provided the contour is chosen so the surface term vanishes. This means

$$e^{-\int t_2}\emptyset(t_2)-e^{-\int t_1}\emptyset(t_1)=0.$$

There are three obvious possibilities (see Fig 7.2). For contours I and II t_1 and $t_2\to\pm i/2$, and the exponential factors are zero. For the closed contour $t_1=t_2$, and the expression vanishes because $\emptyset$ returns to its initial value on traversing the contour.

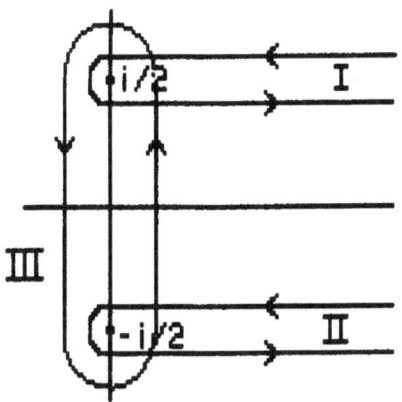

Fig.7.2

The solution regular at
the origin is given by the
contour III: expanding the the
exponential in the integrand
gives a power series expansion
in $\mathfrak{s}$. On the other hand, for
contours I and II one sees
that for small $\mathfrak{s}$ the
contributions to the integral
come from large t. For large t
the integral becomes

$$\int^{\infty} e^{-\mathfrak{s}t} dt/t \approx \int^{1/\mathfrak{s}} dt/t \approx \ln(1/\mathfrak{s}).$$

Thus I and II are irregular solutions. Of course there are
only two linearly independent solutions of the differential
equation. By deforming contour III we see that the regular
solution

$$u = u_{III} = u_I + u_{II}.$$

The solutions u_I and u_{II} are useful for determining the
asymptotic behavior of u for large $\mathfrak{s}$.

To fix the phases of the factors $(t \pm i/2)$ we shall
follow the convention used by Whittaker and Watson[3] in
their chapter on the confluent hypergeometric function. We
rewrite (7.11) as

$$u_C = -(1/2\pi i) \int_C e^{-\mathfrak{s}t} [-(t+i/2)]^{-(1/2)+i\omega} [-(t-i/2)]^{-(1/2)-i\omega} dt$$

(7.12)

(C denoting the contour in question). If $z = t \pm i/2$, we choose
the phase of $-z$ to approach $-\pi$ as z approaches the real
axis from above and to approach π as z approaches the real
axis from below.

If in u_I we put $\tau = \mathfrak{s}(t-i/2)$ and in u_{II} we put
$\tau = \mathfrak{s}(t+i/2)$, (7.12) takes the forms

---

[3] E.T. Whittaker and G.N. Watson, Modern Analysis,
Cambridge (1927).

$$u_I = -(1/2\pi i)e^{-i\zeta/2}(-i\zeta)^{-(1/2)+i\omega}$$

$$\cdot \int_{0^+}^{\infty} e^{-\tau}(-\tau)^{-(1/2)-i\omega}(1-i\tau/\zeta)^{-(1/2)+i\omega}d\tau,$$

$$u_{II} = -(1/2\pi i)e^{i\zeta/2}(i\zeta)^{-(1/2)-i\omega}$$

$$\cdot \int_{0^+}^{\infty} e^{-\tau}(-\tau)^{-(1/2)+i\omega}(1+i\tau/\zeta)^{-(1/2)-i\omega}d\tau,$$

the contour coming in from ∞, circling the origin counter-clockwise, and returning to ∞. By expanding the last factor in the integrand we obtain the asymptotic series for u_I and u_{II} in powers of $1/\zeta$. The corresponding expressions for $v(\eta)$ are obtained by replacing ζ by η and ω by ν.

We see that in the product

$$\psi = uv = (u_I + u_{II})(v_I + v_{II})$$

the asymptotic form will, in general, have four types of term, proportional to

$$e^{(i/2)(\zeta+\eta)}, \quad e^{(i/2)(\zeta-\eta)}, \quad e^{-(i/2)(\zeta+\eta)}, \quad e^{-(i/2)(\zeta-\eta)},$$

i.e. to

$$e^{ipr} \quad , \quad e^{ipz} \quad , \quad e^{-ipr} \quad , \quad e^{-ipz}.$$

For our scattering problem we want only the first two. The unwanted terms are those proportional to $e^{-i\zeta/2}$; they come from $u_I(\zeta)$. The desired boundary condition is achieved if $u_I(\zeta)=0$, which will be true if there is no singularity at $\tau=0$, i.e. if $-(1/2)-i\omega=n$, with n zero or a positive integer. The solution $u_{II}(\zeta)$ is then

$$u_{II} = -(1/2\pi i)e^{i\zeta/2}(i\zeta)^n\int_{0^+}^{\infty} e^{-\tau}(-\tau)^{-(n+1)}(1+i\tau/\zeta)^n d\tau.$$

One readily checks that $v_I(\eta)$ is then proportional to

$$v_I \approx e^{-i\eta/2}(-i\eta)^{n+i\gamma},$$

so the product, which gives the "plane wave" term, is proportional to

$$u_{II}v_I \approx e^{(1/2)i(\zeta-\eta)}(\zeta\eta)^n\eta^{i\gamma} = \rho^{2n}e^{i[pz+\gamma\ln(2pr\sin^2(\theta/2)]}.$$

We thus must take n=0. This choice is akin to the choice $\omega=0$ in the W.K.B. solution. With n=0, u_{II} becomes

$$u_{II}=(1/2\pi i)e^{i\delta/2}\int_{0^+}^{\infty}e^{-\tau}d\tau/\tau=e^{i\delta/2},$$

which is just the same as the W.K.B. result $S_1=\delta/2$. The v_I and v_{II} solutions become

$$v_I=-(1/2\pi i)e^{-i\eta/2}(-i\eta)^{i\gamma}\int_{0^+}^{\infty}e^{-\tau}(-\tau)^{-(1+i\gamma)}(1-i\tau/\eta)^{i\gamma}d\tau,$$

$$v_{II}=-(1/2\pi i)e^{i\eta/2}(i\eta)^{-(1+i\gamma)}\int_{0^+}^{\infty}e^{-\tau}(-\tau)^{i\gamma}(1+i\tau/\eta)^{-(1+i\gamma)}d\tau.$$

The asymptotic forms of v_I and v_{II} are readily found by expanding the integrands in powers of $1/\eta$ and using the relation

$$1/\Gamma(z)=-(1/2\pi i)\int_{0^+}^{\infty}e^{-\tau}(-\tau)^{-z}d\tau.$$

One obtains, to order $1/r$,

$$\psi=u_{II}(\delta)[v_I(\eta)+v_{II}(\eta)]$$

$$\approx\frac{(-i)^{i\gamma}}{\Gamma(1+i\gamma)}\{e^{i(\delta-\eta)/2}\eta^{i\gamma}(1-\frac{i\gamma^2}{\eta})-\gamma e^{i(\delta+\eta)/2}\eta^{-(1+i\gamma)}e^{-2i\eta_0}\}$$

$$\approx\frac{e^{\pi\gamma/2}e^{-i\eta_0}}{|\Gamma(1+i\gamma)|}\{e^{i[pz+\gamma\ln(2pr\sin^2(\theta/2))]}(1-\frac{i\gamma^2}{2pr\sin^2(\theta/2)})$$

$$+f(\theta)e^{i[pr-\gamma\ln(2pr)]}/r\}, \qquad (7.13)$$

with

$$f(\theta)=-(\gamma/2p\sin^2(\theta/2))e^{-i\gamma\ln(\sin^2(\theta/2))+2i\eta_0},$$

and

$$\eta_0=\arg\Gamma(1+i\gamma).$$

This differs from the W.K.B. result only in the definition of η_0. The W.K.B. equation for η_0, (7.08), is the asymptotic form of $\arg\Gamma(1+i\gamma)$ for large γ.[*]

The solution we have found can be expressed in terms of a confluent hypergeometric function. If we renormalize ψ by removing the factor multiplying the curly bracket in (7.13) we have

---

[*] See E. Jahnke and F. Emde, Tables of Functions, Fourth ed. Dover, N.Y. (1945), p.11.

$$\psi = e^{-\pi\gamma/2}\Gamma(1+i\gamma)F(-i\gamma;1;i\eta), \qquad (7.14)$$

where the confluent hypergeometric function, F, is defined
by the series

$$F(a;b;z) = 1 + \frac{a}{b}z + \frac{a(a+1)}{2!b(b+1)}z^2 + \cdots.$$

In terms of the functions defined by Whittaker and Watson,
v_I and v_{II} are related to $W_{(1/2)+i\gamma,0}(i\eta)$ and
$W_{-(1/2)-i\gamma,0}(-i\eta)$ and v to $M_{(1/2)+i\gamma,0}(i\eta)$.

If the Schrodinger equation is separated in polar
coordinates the radial solutions can also be expressed in
terms of confluent hypergeometric functions. The regular
solutions are

$$u_l = e^{-\pi\gamma/2}\frac{|\Gamma(1+l+i\gamma)|}{2(2l+1)!}(2pr)^{l+1}e^{ipr}F(l+1+i\gamma;2l+2;-2ipr)$$

$$\approx \sin[pr - \gamma\ln(2pr) - \pi l/2 + \eta_l], \qquad (7.15)$$

with

$$\eta_l = \arg\Gamma(1+l+i\gamma). \qquad (7.16)$$

We again have that the W.K.B. approximation for η_l, given
by (7.04) is the asymptotic form of the exact relationship,
(7.16), for large γ (see Whittaker and Watson, p. 279).
The irregular solutions, whose asymptotic form is the
cosine of the same argument, can be obtained near r=0 by
means of the relations given by Whittaker and Watson,
p.346. For example, for s waves the regular and irregular
solutions are given, near the origin, by

$$u_0 = |\psi(0)|pr(1+\gamma pr),$$
$$(7.17)$$
$$u_{irreg} = (1/|\psi(0)|)\{1+2\gamma pr[\ln(2pr)+Re\Psi(1+i\gamma)+2C-1]\},$$

where $|\psi(0)|$ is given by (7.18), $\Psi(z)$ is the logarithmic
derivative of the Γ function and $C = -\Psi(1) = .5772\cdots$ is
Euler's constant.

Returning to the complete solution, we see from (7.14)
that the magnitude of the wave function at the origin is
given by

$$|\psi(0)|^2 = e^{-\pi\gamma}|\Gamma(1+i\gamma)|^2 = e^{-\pi\gamma}\pi\gamma/\sinh(\pi\gamma) = 2\pi\gamma/[e^{2\pi\gamma}-1]. \quad (7.18)$$

In the absence of a Coulomb field we would have $|\psi(0)|^2=1$; thus this factor represents the change in $|\psi(0)|^2$ produced by the Coulomb field. At high velocity $\gamma\to 0$, and the factor is unity. As $v\to 0$, $\gamma\to\pm\infty$. For a repulsive Coulomb field $\gamma\to\infty$, and the factor becomes small,

$$|\psi(0)|^2=2\pi\gamma e^{-\pi\gamma},$$

reflecting the difficulty of penetrating the Coulomb barrier. For an attractive field $\gamma\to -\infty$ and the factor is large,

$$|\psi(0)|^2=2\pi(-\gamma).$$

When we are considering a reaction in which the interaction region is within a range R, and $pR\ll 1$, the Born approximation cross section calculated using plane waves for ingoing or outgoing particles can be corrected for the influence of a Coulomb field by allowing for the change in magnitude of the wave functions near the origin. The matrix element is multiplied by $\psi(0)$ or $\psi(0)*$ and the cross section by $|\psi(0)|^2$. For example, for the emission of a charged particle in an attractive Coulomb field, the cross section near threshold is multiplied by $2\pi(-\gamma)$, a factor which becomes infinite at threshold ($v=0$). However this factor just cancels out the phase space factor, p, which appears in the cross section; the factor $p=2\pi/\lambda$ is replaced by $2\pi/a_B$, with $a_B=1/Z_1 Z_2 e^2\mu$ the Bohr radius, and the cross section is finite at threshold. These remarks hold if the particle is emitted in an s state. If it is emitted in a state of angular momentum l, the ratio of $u_l^{Coulomb}$ to u_l^{free} near the origin is, as we see from (7.15),

$$e^{-\pi\gamma/2}|\Gamma(1+l+i\gamma)|/\Gamma(l+1),$$

the value for u_l^{free} being obtained by putting $\gamma=0$. For large negative γ, $-\gamma\gg l+1$, this becomes

$$\surd(2\pi)(-\gamma)^{l+1/2}/\Gamma(l+1),$$

and the square of this quantity multiplies the cross section. However in this case the cross section is

proportional to p^{2l+1} so again the cross section is finite
at threshold. A well known example is the sharp absorption
edges found in the absorption of photons by the emission of
photo-electrons from atoms.

The phase factor by which the exact scattering
scattering amplitude, (7.13) differs from the Born
approximation drops out in the expression for the cross
section. It will have an
effect in situations where
there is interference between
two scattered waves. Such a
situation arises in the
scattering of two
indistinguishable particles,
e.g. the scattering of α's by
α's. In this case the wave
function must be symmetrical
in the coordinates of the two
α-particles, and the

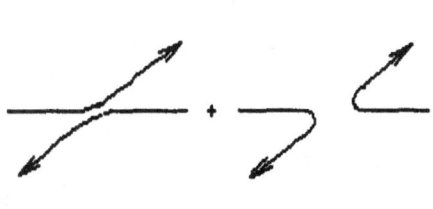

Fig.7.3

amplitudes of the two scatterings illustrated in Fig 7.3
must be added. The cross section is

$$d\sigma = |f(\theta)+f(\pi-\theta)|^2 d\Omega,$$

in contrast to the classical Rutherford cross section,
which is

$$d\sigma = (|f(\theta)|^2+|f(\pi-\theta)|^2) d\Omega.$$

At $\Omega=90°$, $f(\theta)=f(\pi-\theta)$, and the quantum mechanical formula
(known as the Mott formula) gives twice as much scattering
as the Rutherford formula. This prediction was confirmed by
Chadwick[a], who measured the α-α scattering at 45° lab
scattering angle (90° center of mass angle).

With the $f(\theta)$ given by (7.13) the Mott formula becomes

[a] J. Chadwick, Proc. Roy. Soc. <u>128</u>, 114 (1930).

$$d\sigma=\frac{(Ze)^4}{(2\mu v^2)^2}\left|\frac{e^{-i\gamma\ln(\sin^2(\theta/2))}}{\sin^2(\theta/2)}+\frac{e^{-i\gamma\ln(\cos^2(\theta/2))}}{\cos^2(\theta/2)}\right|^2 d\theta$$

$$=\frac{(Ze)^4}{(2\mu v^2)^2}\{\frac{1}{\sin^4(\theta/2)}+\frac{1}{\cos^4(\theta/2)}$$

$$+\frac{2\cos[\gamma\ln(\tan^2(\theta/2))]}{\sin^2(\theta/2)\cos^2(\theta/2)}\}d\Omega.$$

Fig 7.4 shows a comparison of predicted Rutherford and Mott scatterings with the scattering of C^{12} by C^{12} observed at a laboratory energy of 5 Mev,[*] for which the large γ ($\gamma=6.2$) makes the phase oscillations particularly noticeable.

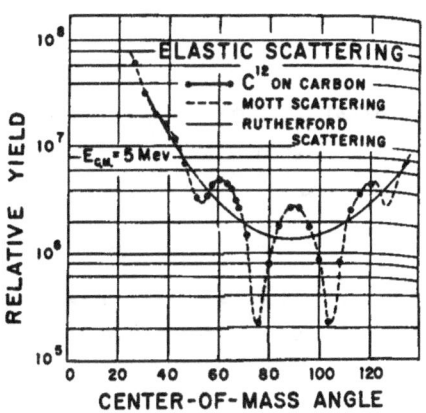

Fig. 7.4

In the case of P-P scattering the orbital wave function must be antisymmetrical if the protons are in the triplet state, symmetrical in the singlet state. For unpolarized protons the scattering is 3/4 triplet, 1/4 singlet, so the Mott formula is

$$d\sigma=(3/4)|f(\theta)-f(\pi-\theta)|^2+(1/4)|f(\theta)+f(\pi-\theta)|^2 d\Omega \qquad (7.19)$$

$$=\frac{e^4}{(2\mu v^2)^2}\{\frac{1}{\sin^4(\theta/2)}+\frac{1}{\cos^4(\theta/2)}-\frac{\cos[\gamma\ln(\tan^2(\theta/2))]}{\sin^2(\theta/2)\cos^2(\theta/2)}\}.$$

---

[*] D.A. Bromley, J.A. Kuehner and E. Almqvist, Phys. Rev. Let. **4**, 365 (1960).

SEC.7II SCATTERING WITH P-P FORCE

We now ask for the change in cross section produced by a spherically symmetric nuclear force. With the inclusion of the nuclear force the wave function is still separable in spherical coordinates, though no longer in parabolic coordinates. The question that then arises is how the Coulomb solution, (7.14), is expressed in terms of the radial functions (7.15), a relationship which must be of the form

$$\psi = \Sigma_l c_l [u_l(pr)/pr] P_l(\mu), \tag{7.20}$$

with $\mu = \cos\theta$.

Let us first consider the simpler case of a free particle, obtained by putting $\gamma = 0$ in (7.14) and (7.15). Equation (7.20) then reads

$$e^{ipz} = \Sigma_l c_l^0 [u_l^0(pr)/pr] P_l(\mu), \tag{7.21}$$

where $u_l^0(pr)$ is (7.15) with $\gamma = 0$. (It can also be expressed in terms of a Bessel function of order $l+1/2$.) Since

$$\int_{-1}^{1} P_l(\mu) P_{l'}(\mu) d\mu = (2/2l+1) \delta_{ll'},$$

we find, by multiplying both sides of (7.21) by $P_{l'}(\mu)$ and integrating,

$$(2/2l+1) c_l^0 u_l^0(pr)/pr = \int_{-1}^{1} e^{ipr\mu} P_l(\mu) d\mu. \tag{7.22}$$

This is an identity in pr, but to determine c_l^0 it is sufficient to consider the relationship for large pr. The left side is then

$$(2/2l+1) c_l^0 \sin(pr - \pi l/2)/pr$$

$$= (1/2l+1) c_l^0 (-i)^l [e^{ipr} - (-1)^l e^{-ipr}]/ipr. \tag{7.23}$$

By a partial integration the right side becomes

$$e^{ipr\mu} P_l(\mu)/ipr \Big|_{-1}^{1} - (1/ipr) \int_{-1}^{1} e^{ipr\mu} (d/d\mu) P_l(\mu) d\mu.$$

The second term is of order $1/(ipr)^2$, as we see by

partially integrating again.[7] To order $1/ipr$ the right
side of (7.22) is thus
$$[e^{ipr}-(-1)^l e^{-ipr}]/ipr,$$
which, equated to (7.23), gives
$$c_l^0=(2l+1)i^l. \qquad (7.24)$$

For the Coulomb case we have, in place of (7.22)

$$(2/2l+1)c_l u_l(pr)/pr=\int_{-1}^{1}\psi(pr,\mu)P_l(\mu)d\mu. \qquad (7.25)$$

However, there is now a difficulty in evaluating the right-
hand side for large pr by partial integration, since the
asymptotic form of $\psi(pr,\mu)$ does not approach a limit as
$\mu \to 1$, as we see from (7.13). Instead, we can evaluate (7.25)
for small pr.

To lowest order in pr we have on the left

$$(2/2l+1)c_l e^{-\pi\gamma/2}|\Gamma(1+l+i\gamma|(2pr)^l/(2l+1)!. \qquad (7.26)$$

The right side is
$$e^{-\pi\gamma/2}\Gamma(1+i\gamma)\int_{-1}^{1}e^{ipr\mu}F(-i\gamma;1;ipr(1-\mu))P_l(\mu)d\mu$$

$$=e^{-\pi\gamma/2}\Gamma(1+i\gamma)\frac{1}{\Gamma(-i\gamma)}\sum_s\frac{\Gamma(-i\gamma+s)}{(s!)^2}(ipr)^s\int_{-1}^{1}(1-\mu)^s e^{ipr\mu}P_l(\mu)d\mu.$$
$$(7.27)$$

The integral can be written as
$$[1-d/d(ipr)]^s\int_{-1}^{1}e^{ipr\mu}P_l(\mu)d\mu,$$

or, from (7.22) and (7.24),
$$2i^l[1-d/d(ipr)]^s u_l^0(pr)/pr=2i^l[1-d/d(ipr)]^s l!(2pr)^l/(2l+1)!,$$

evaluating u_l^0 to lowest order in pr, as we see from (7.15)
with $\gamma=0$. Again, the lowest order term in pr is obtained
by taking the maximum number of derivatives, which gives
$$[l!/(2l+1)!]2^{l+1}i^{l-s}(-1)^s(d^s/d(pr)^s)(pr)^l$$
$$=[l!/(2l+1)!]2^{l+1}i^{l-s}(-1)^s[l!/(l-s)!](pr)^{l-s},$$

---

[7] It may be remarked that the series in $1/ipr$ obtained by
successive partial integrations would terminate with the
l plus first term, checking (7.22) for all pr, not only pr
large.

and substituting this for the integral in (7.27) gives

$$2i^l e^{-\pi\gamma/2}(2pr)^l \frac{1!}{(2l+1)!} \frac{\Gamma(1+i\gamma)}{\Gamma(-i\gamma)} \sum_s (-1)^s \frac{\Gamma(-i\gamma+s)}{(s!)^2} \frac{1!}{(1-s)!} .$$

Equating this to (7.26), we find

$$c_1 = (2l+1)i^l \frac{1!\Gamma(1+i\gamma)}{|\Gamma(1+l+i\gamma)|\Gamma(-i\gamma)} \sum_s (-1)^s \frac{\Gamma(-i\gamma+s)}{(s!)^2} \frac{1!}{(1-s)!} .$$

To put the sum in more recognizable form, we write

$$(-1)^s 1!/(1-s)! = (-1)^s 1(1-1)\cdots(1-s+1)$$

$$= (-1)(-1+1)\cdots(-1+s-1) = \Gamma(-1+s)/\Gamma(-1)$$

and

$$\frac{1}{\Gamma(-i\gamma)} \sum_s (-1)^s \frac{\Gamma(-i\gamma+s)}{(s!)^2} \frac{1!}{(1-s)!} = \frac{1}{\Gamma(-i\gamma)\Gamma(-1)} \sum_s \frac{\Gamma(-i\gamma+s)\Gamma(-1+s)}{(s!)^2}$$

$$= F(-i\gamma,1;1;1),$$

where F is the hypergeometric function. Since (Whittaker and Watson, p.282)

$$F(a,b;c;1) = \Gamma(c)\Gamma(c-a-b)/[\Gamma(c-a)\Gamma(c-b)],$$

we have

$$F(-i\gamma,-1;1;1) = \Gamma(1+l+i\gamma)/[\Gamma(1+i\gamma)1!],$$

and

$$c_1 = (2l+1)i^l \Gamma(1+l+i\gamma)/|\Gamma(1+l+i\gamma)| = (2l+1)i^l e^{i\eta_1}.$$

The Coulomb solution, which we now distinguish with a superscript C, is

$$\psi^C = \Sigma_1 (2l+1)i^l e^{i\eta_1}[u_1^C(pr)/pr]P_1(\mu).$$

Suppose that when the nuclear force is included the asymptotic form of the radial function is changed to

$$u_1(pr) \approx \sin(pr-\gamma\ln(2pr)-\pi l/2+\eta_1+\delta_1),$$

i.e., the nuclear force produces an additional phase shift δ_1. The wave function satisfying the outgoing boundary condition will be

$$\psi = \Sigma_1 (2l+1)i^l e^{i(\eta_1+\delta_1)}[u_1(pr)/pr]P_1(\mu),$$

the phase being chosen so that in the asymptotic form the phase of the ingoing wave is the same as for ψ^C. We then have

$$\psi = \psi^C + \Sigma_1 (2l+1)i^l [e^{i(\eta_1+\delta_1)}u_1(pr) - e^{i\eta_1}u_1^C(pr)]P_1(\mu)/pr,$$

and the asymptotic form is the same as the renormalized

(7.13) with

$$f(\theta)=f^C(\theta)+(1/2ip)\Sigma_1(2l+1)e^{2i\eta_1}(e^{2i\delta_1}-1)P_1(\mu).$$ (7.28)

In the absence of a Coulomb field ($\gamma=0$) this reduces to

$$f(\theta)=(1/2ip)\Sigma_1(2l+1)(e^{2i\delta_1}-1)P_1(\mu),$$ (7.29)

a generalization of (2.08).

If only the s wave scattering is important, (7.28) becomes

$$f(\theta)=e^{2i\eta_0}[\frac{-\gamma}{2p\sin^2(\theta/2)}e^{-i\gamma\ln(\sin^2(\theta/2))}+\frac{1}{2ip}(e^{2i\delta_0}-1)].$$
(7.30)

Note that the additional term due to nuclear scattering will cancel out in the first term of the Mott formula for P-P scattering, (7.19); only odd l terms terms contribute to the first term (triplet scattering), only even l terms to the second term (singlet scattering).

If, for P-P scattering, the nuclear term in (7.30) is larger than the Coulomb term near $\theta=\pi/2$, the scattering will be like one of the two cases shown in Fig 7.5. If the nuclear force is repulsive, like the Coulomb force, there will be constructive interference between the Coulomb and nuclear scattering amplitudes when the two are of comparable magnitude, and we get the upper of the curves shown. For an attractive nuclear force (but one not

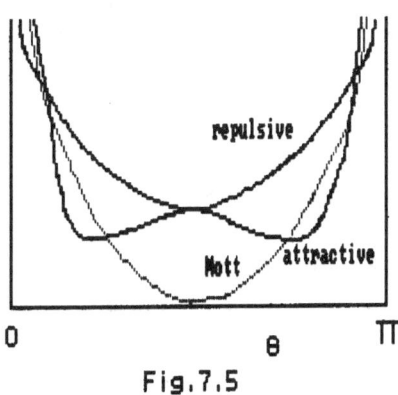

Fig.7.5

strong enough to produce a bound state) there will be destructive interference and we get the lower curve.

White's experimental results showed the forces were attractive, and could be fit with a phase shift δ_0=45°. At the time the potential was represented by a square well with range equal to the classical electron radius. A 45° phase shift required a well depth of V_{P-P}=12.7 Mev,[*] very close to the critical depth (Equation (2.04), $\pi^2/4MR^2$= 12.9 Mev, which (in the absence of a Coulomb field) would just produce a bound state.

The P-P scattering experiment was greatly improved as soon as monoenergetic beams in the 1 Mev range became available from Van de Graaff accelerators. The first such experiments were carried out, the year following White's experiment, by Heydenberg, Hafstad and Tuve.[7] The results were analyzed by Breit, Condon and Present,[10] who obtained, again assuming a square well of range equal to the classical electron radius,

$$V_{P-P}=11.1 \text{ Mev.}$$

The N-P scattering cross section was not well known at the time. Using a value due to Fermi and Amaldi, σ_{N-P}=12 b, Breit, Condon and Present obtained, for the N-P singlet potential,

$$V_{N-P}=11.2 \text{ Mev.}$$

(With the correct value, σ_{N-P}=20.44 b, one would find V_{N-P}=11.75 Mev.)

On the basis of this somewhat accidental very close agreement, Breit, Condon and Present proposed a generalization (the experiments concerned only the ^1S states): "the agreement is so striking as to suggest that the interaction between heavy particles are universally

---

[*] In White's paper there appears to be an error of transposition; the number is incorrectly given as 17.2.
[7] N.P. Heydenberg, L.R. Hafstad and M.A. Tuve, Phys. Rev. 49, 402 (1936); 50, 806 (1936).
[10] G. Breit, E.U. Condon and R.D. Present, Phys. Rev. 50, 825 (1936).

equal, i.e. that the only essential difference between like
and unlike particles is due to the exclusion principle".
The same symmetry of the nuclear forces had earlier been
suggested by Young[11] on the basis of the systematics of
nuclear stability.

It should be noted that Breit, Condon and Present's
statement implies that the like particle forces are equal,
i.e. that the forces between neutrons are the same as those
between protons. The facts of nuclear stability, e.g. the
favoring of even A for odd Z, had already lead to the
conclusion that there were strong attractive N-N forces and
similar P-P forces.[12] Evidence for the equality of the N-N
and P-P forces came from the the observation[13] that the
difference in binding energies of He^3 (two protons and one
neutron) and H^3 (one proton and two neutrons) could be
accounted for solely on the basis of the Coulomb repulsion
between the protons.

Additional evidence for the equality of N-N and P-P
forces had been provided by Fowler, Delsasso and Lauritsen[14]
though somehow this work was overlooked at the time. These

TABLE 7.1		
	ΔE	0.5Z
$B^9 \to Be^9$	2.1	2.0
$C^{11} \to B^{11}$	2.9	2.5
$N^{13} \to C^{13}$	3.0	3.0
$O^{15} \to N^{15}$	3.5	3.5
$F^{17} \to O^{17}$	3.9	4.0

[11] L.A. Young, Phys. Rev. 47, 972 (1935).
[12] K. Guggenheim, J. Phys. 5, 475 (1934); L.A. Young, Phys.
Rev. 47, 972 (1935).
[13] H.A. Bethe, Rev. Mod. Phys. 9, 71 (1937)
[14] W.A. Fowler, L.A. Delsasso and C.C. Lauritsen, Phys.
Rev. 49, 561 (1936).

authors studied the decays by positron emission of a series of light nuclei for which (as for H^3-He^3) the parent and daughter nuclei differed by interchange of proton and neutron numbers. The difference in binding energies they deduced from the observed upper limits of the positron spectra are shown in Table 7.1. It will be seen that they are represented fairly well by

$$\Delta E = 0.5Z, \tag{7.31}$$

where Z is the charge of the daughter nucleus. For Z protons uniformly distributed in a sphere of radius R the Coulomb energy would be

$$E_C = (3/5) Z(Z-1) e^2 / R,$$

and the difference in Coulomb energies between Z-1 and Z would be

$$\Delta E_C = (6/5) Z e^2 / R.$$

This agrees with (7.31) with R=3.4 f. The authors concluded that the N-N and P-P nuclear forces were equal, and that the observed binding energy differences could be attributed to the Coulomb interaction. A more complete discussion of the energy and more modern and extensive results on the binding energies of the "mirror nuclei" will be given in the next section.

SEC.7III THE MIRROR NUCLEI

If we consider the nucleus to be a uniformly charged sphere with charge Ze and radius R the Coulomb energy would be

$$E_C^0 = (3/5) Z^2 e^2 / R. \tag{7.32}$$

At first sight one might be tempted to replace Z^2 by $Z(Z-1)$, since there are $(1/2)Z(Z-1)$ pairs of protons (the interaction of the proton with itself should be excluded). However there is another correction, due to the correlation in the positions of the protons imposed by the exclusion principle. For any potential of the form $V(\underline{r}-\underline{r}')$ the energy is the sum of "ordinary" and "exchange" terms,

$$E_V = (1/2) \Sigma_{nn'} \{ \int |u_n(\underline{r})|^2 |V(\underline{r}-\underline{r}')| u_{n'}(\underline{r}')|^2 d\underline{r} d\underline{r}'$$
$$- \int [u_n(\underline{r})^* u_{n'}(\underline{r})] V(\underline{r}-\underline{r}') [u_{n'}(\underline{r}')^* u_n(\underline{r}')] d\underline{r} d\underline{r}' \}. \tag{7.33}$$

The sums over n and n' are taken over all occupied states, the $u_n(\underline{r})$ are the spinor wave functions, and n includes both the orbital and spin quantum numbers. The brackets in the exchange term show how the spinor products are taken. It should be noted that when n=n' the ordinary and exchange terms cancel; thus the interaction of the proton with itself is eliminated.

Since the proton density is

$$\rho(\underline{r}) = \Sigma_n |u_n(\underline{r})|^2,$$

the ordinary term is just

$$(1/2) \int \rho(\underline{r}) V(\underline{r}-\underline{r}')) \rho(\underline{r}') d\underline{r} d\underline{r}'. \tag{7.34}$$

The effect of the positional correlation depends on whether the range of the interaction is large or small compared to the correlation distance. We can illustrate this by considering two limiting cases of large and small ranges, V=constant and $V=a \, \delta^2(\underline{r}-\underline{r}')$. For V=constant the ordinary term is $(1/2)Z^2 V$, and, since $\int u_n^*(\underline{r}) u_{n'}(\underline{r}) d\underline{r} = \delta_{nn'}$, the exchange term is $-(1/2)ZV$. Thus

$$E = (1/2)Z(Z-1)V, \qquad (7.35)$$

as it evidently should be; the exchange term simply eliminates the self-interaction. For the δ function interaction the ordinary term is $(1/2)a^2\int\rho(\underline{r})^2 d\underline{r}$. The value of the exchange term depends on the degeneracy of the system; since $[u_n^*(\underline{r})u_{n'}(\underline{r})]$ depends on the spin quantum numbers through a factor $\delta_{m_s m_s'}$, in the exchange term a proton only interacts with protons of parallel spin. If the system is fully degenerate, i.e. if each orbital state is doubly occupied, this reduces the exchange term by a factor two, and

$$E_V = (1/2)a^2\int\rho(\underline{r})^2 d\underline{r} - (1/4)a^2\int\rho(\underline{r})^2 d\underline{r}. \qquad (7.36)$$

For the δ function interaction the correlation thus reduces the energy to half the ordinary term, because the exclusion principle prevents a proton being approached by another of the same spin.

If we take as a model a degenerate Fermi gas, i.e. $u_n = e^{i\underline{p}\cdot\underline{r}}\chi_{m_s}$ and all orbital states doubly filled for $p < p_f$, the correlation distance will evidently be of order $1/p_f$. If this is small compared to the nuclear radius the correlation energy is predominantly a volume proportional term, the extent of the non-locality being determined either by the range of the interaction or $1/p_f$, whichever is smaller. The volume term is obtained by writing $d\underline{r}d\underline{r}'$ in the exchange term as $d\underline{r}d(\underline{r}'-\underline{r})$ and extending the range of the $d(\underline{r}'-\underline{r})$ integration to infinity. The exchange term is then

$$E_{exchange} = -\int d\underline{r}\int^{p_f} V_{\underline{p}'-\underline{p}} d\underline{p}d\underline{p}'/(2\pi)^6. \qquad (7.37)$$

where $V_{\underline{p}'-\underline{p}}$ is the Fourier transform of $V(\underline{r})$.

If the range of $V(\underline{r})$ is large compared to $1/p_f$ the contributions to the momentum integrals come from $|\underline{p}'-\underline{p}| << p_f$ and (7.37) becomes, writing $\underline{P} = \underline{p}' - \underline{p}$,

$$E_{exchange} = -\int d\underline{r}\int^{p_f} d\underline{p}/(2\pi)^3\int^{\infty} V_{\underline{P}} d\underline{P}/(2\pi)^3.$$

From
$$V(\underline{r}) = \int^{\infty} V_{\underline{p}} e^{i\underline{P}\cdot\underline{r}} d\underline{P}/(2\pi)^3$$
it follows that
$$V(0) = \int^{\infty} V_{\underline{p}} d\underline{P}/(2\pi)^3,$$
and
$$E_{exchange} = -V(0) \int d\underline{r} (4\pi/3) p_f^3/(2\pi)^3,$$
or, since
$$\rho = 2(4\pi/3) p_f^3/(2\pi)^3, \qquad\qquad (7.38)$$
$$E_{exchange} = -(1/2)V(0)\int \rho d\underline{r} = -(1/2)ZV(0),$$
as in the V=constant case.

In the other limit, if the range of $V(\underline{r})$ is small compared to $1/p_f$, we can replace $V_{\underline{p}'-\underline{p}}$ by V_0 (which is just the volume integral of $V(\underline{r})$) and the exchange energy becomes
$$E_{exchange} = -(1/4)V_0\int \rho^2 d\underline{r}.$$
For $V = a^2\delta(\underline{r}-\underline{r}')$, this agrees with (7.36).

In the Coulomb case
$$V_{\underline{p}'-\underline{p}} = 4\pi e^2/|\underline{p}'-\underline{p}|^2$$
and
$$E_{exchange} = -[4\pi e^2/(2\pi)^6]\int d\underline{r} \int^{p_f} d\underline{p}d\underline{p}'/|\underline{p}'-\underline{p}|^2.$$
It is evident on dimensional grounds that the momentum integral is proportional to p_f^4; an elementary integration gives
$$\int^{p_f} d\underline{p}d\underline{p}'/|\underline{p}'-\underline{p}|^2 = 4\pi^2 p_f^4,$$
and
$$E_{exchange} = -[2e^2/(2\pi)^3]\int p_f^4 d\underline{r} = -(3e^2/4\pi)\int p_f\rho d\underline{r},$$
on using (7.38).

If we take as our nuclear model a sphere of constant density and radius R, the ordinary Coulomb energy is given by (7.32) and the total is
$$E_C = (3/5)Z^2 e^2/R - (3/4\pi)Ze^2 p_f.$$
The proton density is $\rho = 3Z/4\pi R^3$, which, combined with (7.38) gives
$$p_f = (9\pi Z/4)^{1/3}/R.$$
If $R = r_0 A^{1/3}$, we have

$$E_C = (3/5) Z^2 e^2 / (r_0 A^{1/3}) - (3/4\pi)(9\pi/4)^{1/3}(e^2/r_0) Z^{4/3}/A^{1/3}.$$

The difference in Coulomb energies for a pair of mirror nuclei is, to sufficient accuracy,

$$\Delta E_C = \frac{\partial E_C(\langle Z \rangle)}{\partial Z} = \frac{6\langle Z \rangle e^2}{5 r_0} A^{-1/3} - \frac{1}{\pi}(\frac{9\pi}{4})^{1/3} \frac{e^2}{r_0}(\frac{\langle Z \rangle}{A})^{1/3},$$

where $\langle Z \rangle$ is the mean charge of the pair. Since $\langle Z \rangle = (1/2)A$,

$$\Delta E_C = (3/5)(e^2/r_0)[A^{2/3} - (5/3\pi)(9\pi/8)^{1/3}]$$

$$= (3/5)(e^2/r_0)[A^{2/3} - 0.808]. \qquad (7.39)$$

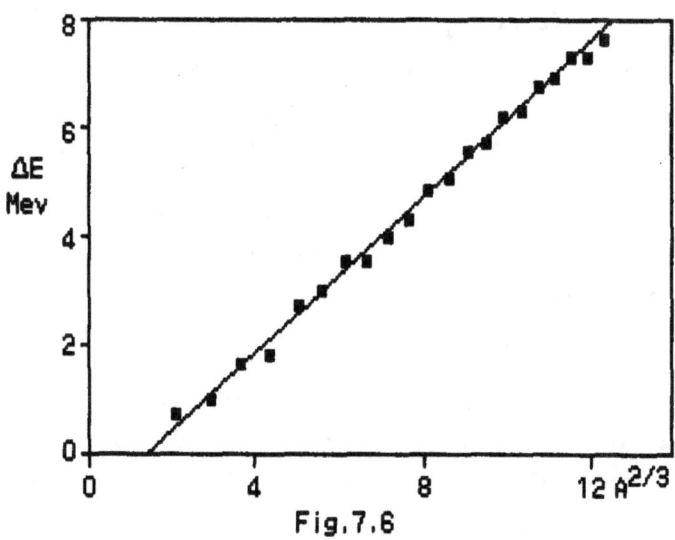

Fig.7.6

Fig 7.6 shows all the measured binding energy differences of the mirror nuclei plotted against $A^{2/3}$, from A=3 to A=43. The results fit a straight line fairly well, aside from a systematic tendency of the points for A=4n+1 to lie a little high with respect to those for 4n+3, which suggests that the A=4n+1 nuclei have slightly smaller radii. The straight line shown in the graph is

$$\Delta E_C = 0.715(A^{2/3} - 1.39) \text{ Mev}.$$

Its slope determines r_0 to be $r_0 = 1.21$ f. The intercept is significantly larger than the value, 0.81 Mev, given by (7.39). This difference can be explained in terms of a somewhat more realistic nuclear model, in which the nuclear edge is rounded rather than cut off sharply, which has been fairly successful in interpreting the results of electron scattering from the heavier nuclei (A>40).

In this model the proton density is taken to be of the form

$$\rho = \rho_0 / [1 + e^{(r-R)/a}]. \tag{7.40}$$

It is supposed that a<<R, so ρ_0 is the interior density. The radius R is that at which the density is reduced by half. These parameters must be consistent with the requirement

$$Z = \int \rho d\underline{r} = 4\pi \int_0^\infty \rho r^2 dr. \tag{7.41}$$

Integrals of the kind appearing in (7.41) can be evaluated by taking advantage of the fact that the gradient of ρ is large only near r=R. To evaluate

$$\int_0^\infty f(r) \rho dr$$

let

$$F(r) = \int_0^r f(r) dr.$$

By a partial integration we get

$$\int_0^\infty f(r) \rho dr = F\rho |_0^\infty - \int_0^\infty F(r) \rho' dr.$$

The surface term vanishes, since $\rho(\infty)=0$ and $F(0)=0$, and in the integral we can expand F(r) in a Taylor series in x=r-R,

$$\int_0^\infty f(r) \rho dr = -\int_{-\infty}^\infty [F(R) + F'(R)x + (1/2)F''(R)x^2 + \ldots] \rho' dx.$$

(Extending the lower limit of integration to $-\infty$ gives an error of order $e^{-R/a}$.) The first term in the bracket gives $\rho_0 F(R)$, the second vanishes because ρ' is an even function of x, and the third, since it is an even function of x, equals

$$-F''(R) \int_0^\infty x^2 \rho' dx,$$

or, by a partial integration,

$$2F''(R)\int_0^\infty x\rho dx=2F''(R)\rho_0\int_0^\infty x/(1+e^{x/a})dx=2F''(R)\rho_0a^2\int_0^\infty y/(1+e^y)dy.$$

With a change of variable to $z=e^{-y}$ we find

$$\int_0^\infty y/(1+e^y)dy=-\int_0^1 \ln z/(1+z)dz=\pi^2/12,$$

according to Pierce,[10] p.64.

We are finally left with

$$\int_0^\infty f(r)\rho dr=\rho_0[\int_0^R f(r)dr+(\pi^2/6)a^2f'(R)+\cdots]. \qquad (7.42)$$

With the help of (7.42) we find that (7.41) gives, to order $(a/R)^2$,

$$Z=(4\pi/3)\rho_0R^3(1+\pi^2(a/R)^2). \qquad (7.43)$$

Equation (7.32) gives the Coulomb energy for a uniformly charged sphere, i.e. one with the density, ρ_1, given by

$$Z=(4\pi/3)\rho_1R^3 \qquad (7.44)$$

for $r\leq R$ and $\rho_1=0$ for $r>R$. The change in Coulomb energy resulting from the change to the density given by (7.40) can be obtained by a simple perturbation calculation. It follows from (7.34), with $V=e^2/|\underline{r}'-\underline{r}|$, that

$$\Delta E_C^0=e\int\emptyset\delta\rho d\underline{r},$$

where $\emptyset$ is the Coulomb potential, $\emptyset(r)=e\int\rho(r')/|\underline{r}'-\underline{r}|d\underline{r}'$. In our case $\delta\rho=\rho-\rho_1$, and

$$\Delta E_C^0=e\int\emptyset(\rho-\rho_1)d\underline{r},$$

with $\emptyset$ the potential due to a uniformly charged sphere, that is

$$\emptyset=(Ze/R)[(3/2)-(1/2)(r/R)^2], \quad r<R,$$

$$\emptyset=Ze/r \qquad\qquad , \quad r>R.$$

However, to the approximation given in (7.42) we can write

$$\Delta E_C^0=(Ze^2/R)\int(\rho-\rho_1)[(3/2)-(1/2)(r/R)^2]d\underline{r},$$

the integral being taken over all space, i.e. we can ignore the change in form of $\emptyset$ at R. The reason is that the

---

[10] B.O. Pierce, A Short Table of Integrals, 2nd Edition, Athenaeum Press, Boston (1910).

correction term in (7.42) depends only on $f'(R)$, and $\emptyset$ and $\emptyset'$ are continuous at $r=R$. Since

$$\int \rho d\underline{r} = \int \rho_1 d\underline{r} = Z,$$

we are left with

$$\Delta E_C^0 = -(Ze^2/2R^3) \int (\rho-\rho_1) r^2 d\underline{r} = -(Z^2 e^2/2R^3)(\langle r^2 \rangle - \langle r_1^2 \rangle),$$

where $\langle r^2 \rangle$ is the mean square radius of the charge distribution

$$\langle r^2 \rangle = \int \rho r^2 d\underline{r} / \int \rho d\underline{r}.$$

The mean square radius for a uniformly charge sphere, $\langle r_1^2 \rangle$, is

$$\langle r_1^2 \rangle = (3/5)R^2,$$

so if we define R_C by

$$\langle r^2 \rangle = (3/5)R_C^2,$$

we have

$$\Delta E_C^0 = -(3/10)(Z^2 e^2/R)[(R_C/R)^2 - 1].$$

In place of (7.32) we then have

$$E_C^0 = (3/5)(Z^2 e^2/R)\{1 - (1/2)[(R_C/R)^2 - 1]\},$$

or, to first order in $(R_C/R)-1$ (second order in a/R),

$$E_C^0 = (3/5)(Z^2 e^2/R)\{1 - [(R_C/R) - 1]\}$$

$$= (3/5)(Z^2 e^2/R)/\{1 + [(R_C/R) - 1]\} = (3/5)Z^2 e^2/R_C. \qquad (7.45)$$

With the aid of (7.42) R_C^2 is easily evaluated; one finds

$$R_C = R[1 + (7/6)\pi^2(a/R)^2]. \qquad (7.46)$$

For the model with the rounded edge (and with the assumption that the neutron density is proportional to the proton density) we have, in analogy to (7.43)

$$A = (4\pi/3)\rho_0^t R^3 [1 + \pi^2(a/R)^2],$$

with ρ_0^t the total nucleon density in the interior. We suppose ρ_0^t independent of A, and define r_0 by $\rho_0^t = 3/(4\pi r_0^3)$. Then

$$R^3 = r_0^3 A[1 - \pi^2(a/R)^2]; \quad R = r_0 A^{1/3}[1 - (\pi^2/3)(a/R)^2].$$

Equation (7.46) becomes

$$R_C = r_0 A^{1/3}[1 + (5\pi^2/6)(a/R)^2],$$

and (7.45) gives

$$E_C^0 = (3/5)[Z^2 e^2/(r_0 A^{1/3})][1 - (5\pi^2/6)a^2/(r_0^2 A^{2/3})].$$

With this change in the ordinary Coulomb term (7.39) becomes

$$\Delta E_C = (3/5)(e^2/r_0)[A^{2/3}-0.808-(5\pi^2/6)(a/r_0)^2],$$

and the intercept at $A^{2/3}=1.39$ is obtained with $a/r_0=0.265$, or, with

$$r_0=1.21 \text{ f, } a=0.32 \text{ f.} \qquad (7.47)$$

The electron scattering experiments of Hofstadter on heavier nuclei gave somewhat different parameters[16]

$$r_0=1.12 \text{ f, } a=0.54 \text{ f.} \qquad (7.48)$$

Another measure of the charge distribution is obtained from the ground state energies of the μ-meson atoms. In fact, the first generally recognized proof that the radius of the charge distribution was significantly smaller than the nuclear radius observed in nucleon-nucleus interactions came from the μ-meson atom experiments of Fitch and Rainwater,[17] who measured the x-ray energies of the 2p-1s transitions. They found energies less than that predicted for a point nucleus, indicating that the 1s level was less tightly bound, a result to be expected because the magnitude of the Coulomb potential of a distributed charge is less than that of a point charge (the 2-p level, since its wave function vanishes at r=0, would be relatively little affected).

Let us suppose that the nuclear charge radius, R_C, is small compared to the Bohr radius, $1/(\alpha Z\mu)$, so $\alpha Z\mu R \ll 1$. Here $1/\mu$ is the meson Compton wavelength. The change in the 1s energy is

$$\delta E_{1s}=\int \psi_{1s}(r)^* V(r)\psi_{1s}(r)d\underline{r},$$

where

$$V=-e(\emptyset-Ze/r),$$

the difference between the actual Coulomb interaction and that for a point charge. For R_C small compared to the Bohr

---

[16] R. Hofstadter, Rev. Mod. Phys. 28, 214 (1956).
[17] V.L. Fitch and J. Rainwater, Phys. Rev. 92, 789 (1953).

adius this becomes
$$\delta E_{1s}=|\psi_{1s}(0)|^2\int V(r)d\underline{r}=(1/\pi)(\alpha Z\mu)^3 V_0,$$
where V_0 is the $\underline{p}=0$ Fourier transform of $V(r)$. The Fourier transform is
$$V_{\underline{p}}=4\pi e^2[(\rho_{\underline{p}}-Z)/p^2.$$
For small $\underline{p}$
$$\rho_{\underline{p}}=\int\rho(r)e^{-i\underline{p}\cdot\underline{r}}d\underline{r}=\int\rho(r)[1-i\underline{p}\cdot\underline{r}-(1/2)(\underline{p}\cdot\underline{r})^2]d\underline{r}$$
$$=Z-(1/6)p^2\int\rho r^2 d\underline{r}=Z[1-(1/6)p^2\langle r^2\rangle].$$
Thus
$$V_0=(2\pi/3)Ze^2\langle r^2\rangle$$
and
$$\Delta E_{1s}=(2/3)(\alpha Z)^4\mu^3\langle r^2\rangle=(2/5)(\alpha Z)^4\mu^3 R_C{}^2.$$
(The ratio of ΔE_{1s} to the unperturbed binding energy, $-E_{1s}=(1/2)(\alpha Z)^2\mu$, is $\Delta E_{1s}/(-E_{1s})=(4/5)(\alpha Z\mu R_C)^2$.)

TABLE 7.2			
	Z	A	r_C
Ti	22	48	1.17
Cu	29	64	1.21
Sb	51	122	1.22
Pb	82	208	1.17

If we write $R_C=r_C A^{1/3}$, the values of r_C found by Fitch and Rainwater are shown in Table 7.2. For the heavy elements more accurate calculations than the first order perturbation treatment were carried out, assuming a uniformly charged sphere. The constants (7.48) agree well with the result for lead and antimony giving $r_C=1.18$ f and $r_C=1.21$ f. The two sets, (7.47) and (7.48), agree for Cu, but give too high a value, $r_C=1.25$ f. For titanium, (7.48) gives $r_C=1.28$ f while (7.47) gives 1.26 f.

SEC.7IV P-P EFFECTIVE RANGE

More modern analyses of the P-P scattering experiments
are in terms of an effective range treatment similar to
that of Sec.2VI. It should first be noted that the phase
shift, δ, produced by a given nuclear potential in the
presence of a Coulomb field is different from the phase
shift, δ^0 which would be produced by the same nuclear
potential in the absence of the Coulomb field. By using the
method of Sec.2VI we can find a relationship between δ and
δ^0, a relationship required if we are to compare the P-P
and N-P scatterings. The derivation depends on the fact
that within the range of the nuclear force the Coulomb
potential is small compared to the nuclear potential. The
latter is of the order of the critical potential, thus we
must have

$$\pi^2/4MR^2 >> e^2/R, \text{ or } R << \pi^2/4e^2M = (\pi^2/8)a_B,$$

where a_B is the Bohr radius for the P-P system,
$a_B = 2/e^2M = 57.60 \text{ f.}$

Let $u(r)$ be the radial solution for the full
potential, $V(r) + e^2/r$, and suppose that asypmtotically, for
large r,

$$u(r) \approx U(r).$$

$U(r)$ is a solution of the Coulomb radial equation; in fact
it is

$$U(r) = |\psi(0)| [\cos\delta u_0^C(r) + \sin\delta u_{irreg}^C(r)]/\sin\delta,$$

where u_0^C and u_{irreg}^C are the regular and irregular Coulomb
solutions, and the normalization is chosen so

$$U(0) = 1,$$

as was done in Sec.2VI. We see from (7.17) that near $r=0$

$$U(r) = 1 + (|\psi(0)|^2 \cot\delta) pr + (2r/a_B) [\ln(2pr) + R\Psi(1 + i\gamma) + 2C - 1]$$

$$(7.49)$$

(remember that $\gamma p = 1/a_B$).

Similarly, let u^0 and U^0 refer to the potential $V(r)$
alone. Near $r=0$

$$U^0 = 1 + prcot\delta^0, \tag{7.50}$$

i.e. it is the same as (7.49) with δ replaced by δ^0, $|\psi(0)|^2 = 1$ and $a_B = \infty$.

The solutions u and u^0 obey the equations

$$u'' + (p^2 - 2\mu V)u = (2/a_B r)u,$$

$$u^{0} {''} + (p^2 - 2\mu V)u^0 = 0.$$

If we multiply the first by u^0, the second by u, and subtract, we obtain

$$(d/dr)(u^0 u' - uu^{0'}) = (2/a_B r)uu^0,$$

whence, since $u(0) = u^0(0) = 0$,

$$u^0(r_0)u'(r_0) - u(r_0)u^{0'}(r_0) = (2/a_B)\int_0^{r_0} uu^0/r\,dr,$$

or

$$\frac{u'(r_0)}{u(r_0)} - \frac{u^{0'}(r_0)}{u^0(r_0)} = \frac{2}{a_B}\int_0^{r_0} \frac{u(r)\ u^0(r)}{u(r_0)u^0(r_0)}\ \frac{dr}{r}\ .$$

If we take r_0 to be the effective range,[10] we may, in the light of our assumption that the Coulomb field is small within the nuclear well, replace $u(r)$ by $u^0(r)$ in the integral. To be definite, suppose V is a square well, of depth near the critical depth. In this case the effective range equals the real range, R, and $u(r_0)$ and $u'(r_0)$ equal $U(r_0)$ and $U'(r_0)$. From (7.49) and (7.50), and neglecting terms of order r_0/a_B, we find

$$|\psi(0)|^2 pcot\delta + (2/a_B)[\ln(2pr_0) + Re\Psi(1 + i\gamma) + 2C] - pcot\delta^0$$

$$= (2/a_B)\int_0^{r_0}[u^0(r)/u^0(r_0)]^2 dr/r. \tag{7.51}$$

For a square well $u^0(r)$ is proportional to $\sin(pr)$ and, if the depth is near the critical depth, the phase at the edge is near $\pi/2$. The right side of the equation then equals

$$(2/a_B)\int_0^{\pi/2}\sin^2 x\,dx/x = (1/a_B)\int_0^{\pi}(1 - \cos x)dx/x = (1/a_B)[C + \ln\pi - Ci(\pi)]$$

---

[10] We take the scattering length and the effective range to be those of (2.36), i.e. those for the nuclear potential only, with no Coulomb field.

(see Jahnke-Emde, p.3). Here Ci is the cosine-integral function, and $Ci(\pi)=0.0737$, so the right side of (7.51) equals $0.824(2/a_B)$. Remembering that $p=1/\gamma a_B$, we have

$$|\psi(0)|^2 p\cot\delta + (2/a_B)[\ln(2r_0/\gamma a_B) + \text{Re}\Psi(1+i\gamma) + 0.330] = p\cot\delta^0,$$

or, using the effective range relation, (2.36), for $p\cot\delta^0$,

$$|\psi(0)|^2 p\cot\delta + (2/a_B)[\ln(1/\gamma) + \text{Re}\Psi(1+i\gamma)]$$

$$= -(1/a) + (2/a_B)[\ln(a_B/2r_0) - 0.330] + (1/2)r_0 p^2, \quad (7.52)$$

It is evident that the above derivation does not depend critically on the potential being a square well; at most, for different shaped potentials, one might expect some change in the constant 0.330.

Later P-P scattering experiments were carried out by Heydenburg, Hafstad and Tuve[19] in the energy range 670 to 870 Kev, and by Herb, Kerst, Parkinson and Plain,[20] who extended the range to 2.4 Mev. If the experimentally determined values of δ are used to plot the left side of (7.52) against p^2 the result is the straight line shown in Fig.7.7, which is copied from Blatt and Weisskopf[21] (the ordinate, K, is actually $a_B/2$ times the left side). From the slope and intercept of the line one determines r_0 and a. The results given by Jackson and Blatt[22] were $a=-17.0\pm0.4$, $r_0=2.67\pm0.07$ f. More recent treatments take into account the fact that the Coulomb potential itself is not exact, but is modified by the vacuum polarization. That this has a significant effect was pointed out by Foldy and Ericksen;[23] it leads to a modification of the right side of

---

[19] N.P. Heydenburg, L.R. Hafstad and M.A. Tuve, Phys. Rev. 56, 1078 (1939)
[20] R.G. Herb, D.W. Kerst, D.B. Parkinson and G.J. Plain, Phys. Rev. 55, 998 (1939)
[21] J.M. Blatt and V.F. Weisskopf, Theoretical Nuclear Physics, Wiley (1952), p.92.
[22] J.D. Jackson and J.M Blatt, Rev. Mod Phys. 22, 77 (1950).
[23] L.L. Foldy and E. Eriksen, Phys. Rev. 98, 755 (1955)

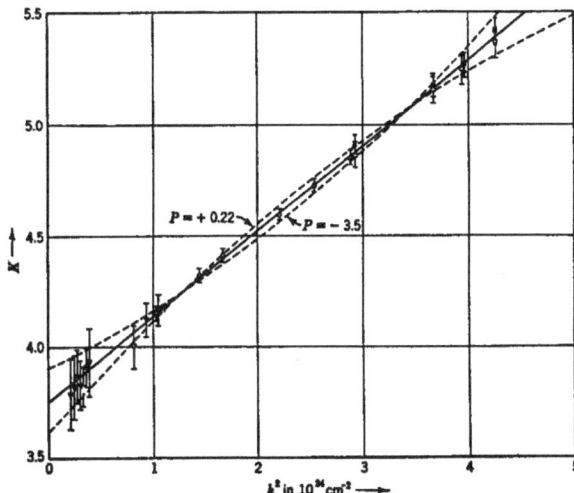

Fig.7.7

(7.52) (see E. Henley, Isospin in Nuclear Physics,[24] p.29).
Henley gives as the best result for P-P scattering

$$a_{PP}=-17.0\pm0.2 \text{ f}, \quad r_{PP}=2.83\pm0.03 \text{ f}.$$

Curiously enough, r_{PP} turns out to be virtually the same as
the range assumed in the original treatments of P-P
scattering.

There are a number of experiments from which an N-N
scattering length can be extracted, though with some
theoretical uncertainty. These involve reactions in which
there are three particles in the final state, two of them
being neutrons. The probability of emitting two neutrons of
small relative momentum is enhanced by their interaction,
an effect already discussed in Sec.4II in connection with
the magnetic dipole photoeffect of the deuteron. As one can
see from (4.12), the differential spectrum is modified by
the momentum dependent factor

$$\sin^2\delta_{NN}/p^2=1/\{p^2+[(1/a_{NN})-(1/2)r_{NN}p^2]^2\},$$

---

[24] Isospin in Nuclear Physics, D.H. Wilkonson, Editor,
North Holland Publishing Co. (1969).

where p is the relative momentum of the neutrons.

The first such reaction to be studied was $\pi^- + D \rightarrow \gamma + 2N$. Panofsky, Aamodt and Hadley[25] measured the γ-ray spectrum and observed a pronounced peaking towards high energies as compared to a phase-space spectrum. They pointed out that this indicated a nearly bound N-N state. A more complete experiment by Haddock, Salter, Heller, Czirr and Nygren[24] detected all three final state particles. It lead to a value $a_{NN} = -18.4 \pm 1.5$ f.

The reaction $D + H^3 \rightarrow He^3 + 2N$ was studied by Baumgarten, Conzlett, Shield and Slobodrian[27] and gave $a_{NN} = -16.1 \pm 2$ f. (a companion experiment on the reaction $D + He^3 \rightarrow H^3 + 2P$ gave a value of a_{PP} which agreed within the errors with the directly measured one) and a rough value of the effective range, $r_{NN} = 3.2 \pm 1.6$ f.

The weighted mean of these two results is (Henley, loc. cit.)

$$a_{NN} = -17.6 \pm 1.5 \text{ f}, \quad r_{NN} = 3.2 \pm 1.6 \text{ f}.$$

The conclusion drawn from the mirror nuclei are thus substantiated for the ^1S state: the N-N and P-P parameters agree. For the N-P system

$$a_{NP} = -23.715 \pm 0.13 \text{ f}, \quad r_{NP} = 2.76 \pm 0.07 \text{ f},$$

and there is a definite difference between a_{PP} and a_{NP}. In terms of well depth, the N-P depth appears to be about 2% deeper than the P-P. The question is whether the observed difference can be ascribed to electromagnetic, rather than strong interaction, effects.

The effects on the scattering lengths of the nuclear magnetic moment interactions and of the N-P mass difference

---

[25] W.K.H. Panofsky, R.L. Aamodt and J. Hadley, Phys. Rev. 81, 565 (1951)
[24] R.P. Haddock, R.M. Salter, Jr, M. Heller, J.B. Czirr and D.R. Nygren, Phys. Rev. Letters 14, 318 (1965); D.R. Nygren, Thesis, Univ. of Washington, unpublished.
[27] E. Baumgarten, H.E. Conzlett, E. Shield and J.R. Slobodrian, Phys. Rev. Letters 16, 105 (1966)

(which in this context is deemed an electromagnetic effect)
can be calculated, and are found to be an order of
magnitude too small to account for the difference between
a_{NP} and a_{PP}. A difference between a_{NN} and a_{PP} is also
deduced, but one much smaller than the probable error of
the measurement. Another indirect electromagnetic effect
comes from the nearly 5 Mev difference between the $\pi^{\pm}$ and
π^0 masses. For the like particles the one-meson exchange
forces, which dominate at large separations, come entirely
from π^0 exchange. For the N-P case there is also a
contribution from charged meson exchange. The effect of
increasing the mass of the charged π's is to give this part
of the potential a shorter range, and also to make it
deeper. The latter effect arises because, if the mesons and
nucleons interact through a relativistic pseudoscalar
interaction, the source term in (6.17) is proportional to
$1/M$ rather than $1/\mu$, where M and μ are the nucleon and
meson masses. The change results in the potential, (6.19),
being multiplied by a factor $(\mu/M)^2$. The critical parameter
of (2.05), $f^2=(g^2/4\pi)(M/\mu)$, is thus proportional to μ/M.
The change in range accounts for half the observed
difference between r_{NP} and r_{PP}. The change in f^2 also turns
out to account for half the difference between a_{NP} and a_{PP}.
It is thus easy to believe that similar corrections for the
shorter range components of the interaction, plus other
small electromagnetic effects, can account for the entire
observed difference.

CHAPTER 8

ISOTOPIC SPIN INVARIANCE

SEC.8I SYMMETRY OF THE NUCLEAR FORCES

The nuclear Hamiltonian, excluding electromagnetic terms, can be written in the second-quantization formalism

$$H=\int \psi_P^+(-1/2M_P)\Delta\psi_P d\underline{r}+\int \psi_N^+(-1/2M_N)\Delta\psi_N d\underline{r}$$
$$+(1/2)\int [\psi_P^+(\underline{r})\psi_P(\underline{r})]V_{PP}(\underline{r}-\underline{r}')[\psi_P^+(\underline{r}')\psi_P(\underline{r}')]d\underline{r}d\underline{r}'$$
$$+(1/2)\int [\psi_N^+(\underline{r})\psi_N(\underline{r})]V_{NN}(\underline{r}-\underline{r}')[\psi_N^+(\underline{r}')\psi_N(\underline{r}')]d\underline{r}d\underline{r}'$$
$$+\int [\psi_P^+(\underline{r})\psi_P(\underline{r})]V_{PN}(\underline{r}-\underline{r}')[\psi_N^+(\underline{r}')\psi_N(\underline{r}')]d\underline{r}d\underline{r}'. \qquad (8.01)$$

While it has not been expressed explicitly, the V's may contain spin-dependent factors and exchange operators.

The symmetry exemplified by the mirror nuclei, sometimes called charge symmetry, is obtained by putting $M_P=M_N$ and $V_{PP}=V_{NN}$. The Hamiltonian is then invariant to interchanging ψ_P and ψ_N, i.e. to introducing new variables

$$\psi_P'=\psi_N, \quad \psi_N'=\psi_P.$$

If, in addition, we put $V_{PN}=V_{PP}=V_{NN}$, as suggested by the measurements on the ^1S states, we have a wider symmetry, sometimes called charge independence. The Hamiltonian then depends on the fields only in the combination

$$\psi_P^+\psi_P+\psi_N^+\psi_N$$

(aside from coordinate and spin dependent operators) and is
invariant to any coordinate and spin independent
transformation which leaves this form invariant.

Mathematically, this problem is identical to the
familiar problem of introducing the two spin-states of the
electron in non-relativistic atomic theory. The group of
transformations in question (called the U_2 group) consists
of the unitary transformations

$$\psi_P'=a\psi_P+b\psi_N, \qquad \psi_N'=c\psi_P+d\psi_N,$$

with

$$S=\begin{vmatrix} a & b \\ c & d \end{vmatrix}$$

a unitary matrix. The matrix S depends on eight parameters,
the real and imaginary parts of four complex numbers. The
unitary condition, $S^+S=1$, gives four conditions and reduces
the number of independent parameters to four. Since S is
unitary Det S has unit magnitude; if

$$\text{Det } S=e^{2i\gamma},$$

we can write

$$S=e^{i\gamma}D,$$

where the matrix D has

$$\text{Det } D=1.$$

Thus the U_2 group factors into the product of the
transformations

$$U_1=e^{i\gamma}\underline{1}$$

and the three parameter group of unitary transformations
with unit determinant (called the SU_2 group).
The U_1 group induces the transformation

$$\psi_P'=e^{i\gamma}\psi_P, \qquad \psi_N'=e^{i\gamma}\psi_N.$$

The Hamiltonian is invariant under this transformation
because every term contains an equal number of ψ's and
ψ^+'s, that is, an equal number of annihilation and creation
operators. The corresponding constant of the motion is the
number of nucleons, which is conserved for the Hamiltonian
(8.01) (which conserves the number of protons and neutrons

separately) and also if electromagnetic and β-decay interactions are included. The remaining SU_2 group is the one of interest to us in connection with the equality of the nuclear forces.

The matrices D can be parameterized, as in the electron spin case, as

$$D=\begin{vmatrix} \alpha* & \beta* \\ -\beta* & \alpha* \end{vmatrix},$$

with α and β the Cayely-Klein parameters,

$$\alpha=\cos(\theta/2)e^{(i/2)(\psi+\phi)}, \quad \beta=\sin(\theta/2)e^{(i/2)(\psi-\phi)}.$$

As in the case of ordinary spin, we can introduce a spinor notation,

$$\psi=\begin{vmatrix} \psi_P \\ \psi_N \end{vmatrix},$$

and the so-called isotopic-spin operators, which we denote by t_x, t_y and t_z to distinguish them from s_x, s_y and s_z,

$$t_z=(1/2)\begin{vmatrix} 1 & 0 \\ 0 & -1 \end{vmatrix}, \quad t_x=(1/2)\begin{vmatrix} 0 & 1 \\ 1 & 0 \end{vmatrix}, \quad t_y=(1/2)\begin{vmatrix} 0 & -i \\ i & 0 \end{vmatrix}$$

The transformations D are then related to rotations in a real three dimensional space (isotopic spin space), with ψ, θ and ∅ the Euler angles of the rotation, through the observation that $\psi^+ t_x \psi$, $\psi^+ t_y \psi$ and $\psi^+ t_z \psi$ transform like the components of a vector in the real space when ψ^+ and ψ are transformed by D.

Also, as for ordinary spin, one can write (8.02) in terms of the unit spinors χ_{m_t}, with

$$\chi_{1/2}=\begin{vmatrix} 1 \\ 0 \end{vmatrix}, \quad \chi_{-1/2}=\begin{vmatrix} 0 \\ 1 \end{vmatrix}.$$

A proton has $m_t=1/2$, a neutron has $m_t=-1/2$. The spinor ψ can be expanded in the form

$$\psi=\Sigma_{nm_s m_t} a_{nm_s m_t} u_n(r)\chi_{m_s}\chi_{m_t},$$

that is, in orbital times spin times isospin functions.

We have implicitly assumed that the ψ's and ψ^+'s in (8.01) satisfy anticommutation relations, e.g. not only $\psi_P(r)\psi_P(r')+\psi_P(r')\psi_P(r)=0$ but also $\psi_P(r)\psi_N(r')+\psi_N(r')\psi_P(r)$

=0. Otherwise the commutation relations would not be
invariant under the transformation D. This means that the
wave functions in our second quantization representation are
antisymmetric under interchange of any two states $n_1 m_{s_1} m_{t_1}$
and $n_2 m_{s_2} m_{t_2}$, or, in a coordinate representation, are
antisymmetric under coordinate, spin and isospin
interchange. If $m_{t_1} = m_{t_2}$ the wave function is antisymmetric
under interchange of $n_1 m_{s_1}$ and $n_2 m_{s_2}$, which is equivalent,
in a coordinate representation, to taking the wave function
antisymmetric in orbit and spins of protons and neutrons
separately. The additional antisymmetry in exchanging
states of different m_t has no consequences in evaluating
the matrix elements of an operator of the form (8.01),
since there are no exchange terms between states of
different m_t; the corresponding statement for the m_s case
has been remarked in Sec.7III.

Historically, the central N-P potential was written

$$V = V_W(|r_1 - r_2|) + V_B(|r_1 - r_2|)(\sigma_1 \cdot \sigma_2) + V_M(|r_1 - r_2|)P_{12}^X$$
$$+ V_H(|r_1 - r_2|)P_{12}^X P_{12}^\sigma, \qquad (8.03)$$

where P^X and P^σ are the coordinate and spin exchange
operators. The four kinds of terms were called Wigner,
Bartlett, Majorana and Heisenberg forces. The spin exchange
operator is

$$P_{12}^\sigma = (1/2)(1 + 4 s_1 \cdot s_2), \qquad (8.04)$$

as can easily be seen in the representation in which
$S = s_1 + s_2$ is diagonalized.[1] Thus (8.03) can be rewritten in
the form

---

[1] The triplet states are symmetric in the spins and have
$P_{12}^\sigma = 1$, the singlet state is antisymmetric and has $P_{12}^\sigma = -1$.
The relationship (8.04) has been chosen to give P_{12}^σ the
correct values in the two cases, since $s_1 \cdot s_2 = 1/4$ for $S=1$
and $s_1 \cdot s_2 = -3/4$ for $S=0$.

$$V=V_1(\underline{r}_1-\underline{r}_2)+V_2(\underline{r}_1-\underline{r}_2)\underline{s}_1\cdot\underline{s}_2+[V_3(\underline{r}_1-\underline{r}_2)+V_4(\underline{r}_1-\underline{r}_2)\underline{s}_1\cdot\underline{s}_2]P^x_{12}.$$

$$(8.05)$$

For the states of the two-particle system $P^x_{12}=1$ for l even, $P^x_{12}=-1$ for l odd. Thus the four types of interaction allow us to assign independent interactions to the four types of states, $S=0$ or 1, l even or odd.

For the N-N and P-P interactions there is no purpose in introducing P^x_{12}; since the wave function, Ψ, is antisymmetric in the two particles,

$$P^x_{12}P^\sigma_{12}\Psi=-\Psi,$$

and multiplying by P^σ_{12} we have

$$P^x_{12}\Psi=-P^\sigma_{12}\Psi=-(1/2)(1+4\underline{s}_1\cdot\underline{s}_2)\Psi. \qquad (8.06)$$

Using this relation the P^x_{12} dependent terms of (8.05) can be reduced to the form of the first two terms.[2] For the two particle system the two types of term allow independent interactions for the two types of state permitted by the exclusion principle, $S=0$ and l even or $S=1$ and l odd.

For a wave function antisymmetric in coordinate, spin and isotopic spin variables we have

$$P^x_{12}P^\sigma_{12}P^\tau_{12}\Psi=-\Psi,$$

and, in place of (8.06)

$$P^x_{12}\Psi=-P^\sigma_{12}P^\tau_{12}\Psi=-(1/4)(1+4\underline{s}_1\cdot\underline{s}_2)(1+4\underline{t}_1\cdot\underline{t}_2)\Psi. \qquad (8.07)$$

Then (8.03) or (8.05) can be written in the form

$$V=V_a(\underline{r}_1-\underline{r}_2)+V_b(\underline{r}_1-\underline{r}_2)(\underline{s}_1\cdot\underline{s}_2)+V_c(\underline{r}_1-\underline{r}_2)(\underline{t}_1\cdot\underline{t}_2)$$
$$+V_d(\underline{r}_1-\underline{r}_2)(\underline{s}_1\cdot\underline{s}_2)(\underline{t}_1\cdot\underline{t}_2) \qquad (8.08)$$

Note that, in the isotopic spin formalism, (8.08) applies to any pair of nucleons, irrespective of whether they are protons or neutrons (which they are is determined

---

[2] Note that $(\underline{s}_1\cdot\underline{s}_2)^2=(3/16)-(1/2)(\underline{s}_1\cdot\underline{s}_2)$, a relation which follows from the fact that, since the eigenvalues of $\underline{s}_1\cdot\underline{s}_2$ are $1/4$ and $-3/4$, $(\underline{s}_1\cdot\underline{s}_2-1/4)(\underline{s}_1\cdot\underline{s}_2+3/4)=0$.

by the wave function on which V operates). As written,
(8.08) exhibits the charge independence, or invariance
under isotopic rotations, which we have been discussing;
this because it depends on isotopic spin only through the
rotationally invariant factor $(\underline{t}_1 \cdot \underline{t}_2)$.

The isotopic spin formalism was used by Heisenberg in
his 1932 paper, and the transformation of the form of
(8.03) by the use of (8.07) was pointed out in 1936 by
Cassen and Condon,[3] but neither Heisenberg nor Cassen and
Condon supposed the forces were charge independent.

In Chapt.6 we derived nuclear forces for the neutral
meson pseudoscalar theory. To make the theory charge
independent (symmetrical pseudoscalar theory) we suppose
there are three real meson fields, $\underline{\emptyset} = (\emptyset_x, \emptyset_y, \emptyset_z)$, which
transform like a vector under isotopic spin rotations
(the three charge eigenfunctions corresponding to $\pi^{\pm}$ and
π^0). The left side of (6.17) becomes a vector under isospin
rotations; to make the right side a vector also we
multiply the source term by $\tau = 2\underline{t}$ (in analogy to $\underline{\sigma} = 2\underline{s}$).
Equations (6.18) and (6.19) are changed only by
multiplication by a factor $\tau_1 \cdot \tau_2$ and replacement of $g_N g_P$ by
g^2, so they become

$$V_T = \frac{g^2}{4\pi} \frac{\tau_1 \cdot \tau_2}{3\mu^2} [1 + \frac{3}{\mu r} + \frac{3}{(\mu r)^2}] \frac{e^{-\mu r}}{r} [\frac{3(\underline{\sigma}_1 \cdot \underline{r})(\underline{\sigma} \cdot \underline{r}_2) - (\underline{\sigma}_1 \cdot \underline{\sigma}_2) r^2}{r^2}],$$

$$V_S = \frac{g^2}{4\pi} \frac{(\tau_1 \cdot \tau_2)(\underline{\sigma}_1 \cdot \underline{\sigma}_2)}{3} [\frac{e^{-\mu r}}{r} + \frac{\delta(\underline{r})}{\mu^2}]. \qquad (8.09)$$

The choice of a source term proportional to τ implies
$g_N = -g_P$; the minus sign is contained in the $\tau_{z1} \tau_{z2}$ term,
which gives the contribution of neutral mesons to $\tau_1 \cdot \tau_2$.
The result for a charged pseudoscalar theory is obtained by
subtracting the contribution of the neutrals, i.e. it is
$\tau_1 \cdot \tau_2 - \tau_{z1} \tau_{z2}$.

---

[3] B. Cassen and E.U. Condon, Phys. Rev. 50, 846 (1936)

For central forces (8.08) would be invariant under separate rotations of the coordinates, spins and isospins, which would have the consequence that the total angular momentum of a nucleon system, $\underline{L}=\Sigma_{i=1}^{A} \underline{l}_i$, would be a constant of the motion, as would the total spin, $\underline{S}=\Sigma \underline{s}_i$ and the total isospin, $\underline{T}=\Sigma \underline{t}_i$, and the states could be characterized by the quantum numbers L, M_L;S,M_S;T, M_T. If we add tensor and spin-orbit forces the good quantum numbers become J,M_J;T,M_T, with $\underline{J}=\underline{L}+\underline{S}$ the total angular momentum.

For the two nucleon system S and T can have the values 0 or 1. The antismmetry of the total wave function requires, for even orbital wave function (l even), either S=1, T=0 or S=0, T=1. (The potential (8.08 in the first case has $\underline{s}_1 \cdot \underline{s}_2=1/4$, $\underline{\tau}_1 \cdot \underline{\tau}_2=-3/4$ and in the second $\underline{s}_1 \cdot \underline{s}_2=-3/4$, $\underline{\tau}_1 \cdot \underline{\tau}_2=1/4$.) For odd l we must have S=0, T=0 or S=1, T=1. The four types of term in (8.08) again allow us to give independent interactions to the four types of states of the N-P system. However the like particle forces are then determined: they are the same, for the states allowed by the exclusion principle, as the N-P forces in the same states; this because the the potentials given by (8.08) are independent of M_T, and so are the same for the three T=1 states. Note that M_T=1 means P-P, M_T=0 means N-P and M_T=-1 is N-N. The T=0 state has M_T=0, and thus refers only to the N-P system.

SEC.8II SELECTION RULES AND MULTIPLETS

If the isotopic spin symmetry were exact the total isotopic spin, $\underline{T}$, of a nuclear system would be a constant of the motion. For example, a resonance state of given T and M_T could only decay into products in a state of the same T and M_T. The symmetry is no longer exact when electromagnetic forces are included; however M_T remains a constant of the motion, since the conservation of M_T is equivalent to the conservation of charge. The charge on a nucleon can be written

$$e_i = (t_{zi} + 1/2)e, \tag{8.10}$$

and the total charge is

$$Ze = \Sigma_i e_i = \Sigma_i (t_{zi} + 1/2)e = (T_z + A/2)e = (M_T + A/2)e. \tag{8.11}$$

Thus the Hamiltonian remains invariant for rotations about the isotopic z axis.

However, though the T selection rule is no longer exact, we may expect it to be nearly valid in view of the relative weakness of electromagnetic forces, particularly for the low lying states of light nuclei, which are not too much affected by the Coulomb interaction.

The first example of this selection rule was given by Oppenheimer and Serber.[*] When B^{11} is bombarded by protons a resonance level is found at a proton energy of 0.16 Mev. This level of the combined nucleus C^{12}, at an excitation energy of 16.1 Mev, was observed to decay with the emission of long range α particles (decay to the ground state of Be^8) and also a 16 Mev γ corresponding to decay to the ground state of C^{12}. The yield of long range α's was observed to be only about ten times the yield of γ's; Oppenheimer and Serber asserted that this could only be understood if a strong selection rule were inhibiting the α

[*] J.R. Oppenheimer and R. Serber, Phys. Rev. 53, 636 (1938)

emission. A crude estimate of the effect to be expected was given by observing that a Coulomb matrix element resonsible fot the T violation would be proportional to e^2, giving a decay rate proportional to e^4, i.e. the square of the fine structure constant, suggesting a reduction by a factor 20000.[a]

They said that the resonance level must be J=2, T=1, an assignment which would allow γ decay to the C^{12} ground state but forbid decay to $Be^8+\alpha$, since both Be^8 and α have T=0. The width of the resonance level is now known to be 6.5 Kev, most of which is due to emission of shorter range α's to a J=2, T=0 level of Be^8 at 2.9 Mev (the width for long range α emission is only 290 ev). The effectiveness of the T selection rule can be seen by comparing this with the decay of a lower energy resonance level of C^{12} for which the α emission is allowed: the J=0, T=0 state at 10.3 Mev decays to the ground state of Be^8 (also J=0, T=0) plus α with a width of 3 Mev, five hundred times wider than the state giving the $\Delta T=1$ transition.

The isospin symmetry would also require a (2T+1)-fold degeneracy for states of given T, i.e. the energy would be independent of M_T. According to (8.11), this means the same energy for a set of states in nuclei of the same A but differing Z. States related in this way are called analogue states. The Coulomb energy, of course, depends on Z and lifts this degeneracy. In the present language, the mirror nuclei discussed in Chapt.7 would be described as the $M_T=\pm1/2$ levels of a T=1/2 state.

The first example of an analogue state illustrating the full symmetry was given by Oppenheimer and Serber in the above quoted paper. Having identified the 16.1 Mev

---

[a]   Perhaps a better estimate would be mutiplication by a factor

$$[(e^2/R)/(\pi^2/2MR^2)]^2\approx1/2000$$

for C^{12} with $R=1.4A^{1/3}$.

level of C^{12} as T=1, M_T=0, they said the analogue M_T=-1 state would be the ground state of B^{12}. This turned out to be not quite correct; the analogue is actually the first exited state, at 0.95 Mev.

Oppenheimer and Serber credit these ideas of the consequences of isospin symmetry to Gregory Breit; their paper found illustrations of Professor Breit's suggestions. The 1937 paper appears to have been forgotten and analogue states were only rediscovered in 1952, when the subject was developed by Adair.

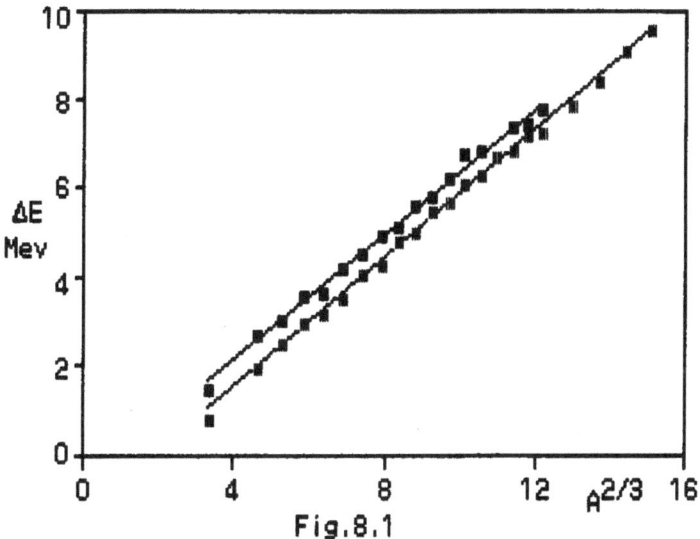

Fig.8.1

Many examples of analogue states are now known. Fig 8.1 is a plot similar to Fig 7.6 for the mirror nuclei; it shows the measured binding energy differences of the lowest T=1 levels for even A nuclei from A=6 to A=58. There are now two points for each A, corresponding to the M_T=1 minus M_T=0 and the M_T=0 minus M_T=-1 differences. The two lines give the Coulomb energy differences, calculated from the formula derived in Sec.7III,

$$E_{Coul} = \frac{3}{5} \frac{e^2}{r_0 A^{1/3}} \{Z^2 [1 - \frac{5\pi^2}{6} (\frac{a}{r_0 A^{1/3}})^2] - \frac{5}{4\pi} (\frac{9\pi}{4})^{1/3} Z^{4/3}\}.$$

The values given by (7.47), r_0=1.21 f, a=0.32 f, derived the mirror nuclei results, have been used in plotting these lines.

CHAPTER 9

HIGHER ENERGY REACTIONS

SEC.9I NUCLEON-NUCLEON SCATTERING

With the completion of the 184'' cyclotron in Berkeley
in 1947 the energies of particle beams available for
experiment jumped an order of magnitude, from tens to
hundreds of Mev's. For the first time, in nucleon-nucleon
scattering, one was dealing with wavelengths short enough
to probe the shape of nuclear forces.

The first high energy nucleon-nucleon scattering
experiment was done by Segrè and his collaborators[1] in 1949.
A 90 Mev neutron beam was produced by stripping the protons
from the 180 Mev deuteron beam in the cyclotron (see
Appendix B) and the scattering of these neutrons by protons
in a paraffin target was measured by observing the recoil
protons. The results are shown in Fig 9.1 (the upper curve
is for a neutron energy of 40 Mev, the lower for 90 Mev).

---

[1] J. Hadley, E.L. Kelley, C. Leith, E. Segrè, C. Wiegand
and H. York, Phys. Rev. 75. 351 (1949).

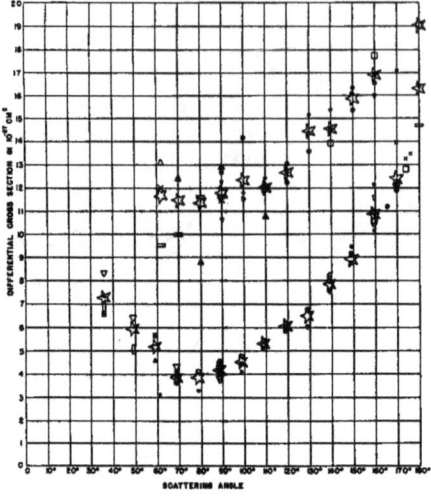

Fig.9.1

The most striking feature of the curve is the
approximate symmetry about 90°. Since asymmetric terms
would arise from interference of waves even and odd l, and
since it is known there is l=0 scattering, exact symmetry
would require no scattering in odd l states. Thus the odd
state forces would have to be zero, which requires, in the
notation of (8.05), $V_1=V_3$ and $V_2=V_4$, so the interaction is
proportional to $(1+P^x_{12})$. Since $P^x_{12}=(-1)^l$, this gives the
desired property. Such an interaction is called a Serber
force and, at the time, the conclusion that the forces have
such a character was a surprising one. The existence of the
repulsive core was not known, and it was generally
believed, because of the argument given in Sec.1I, that the
linear dependence of nuclear binding energy on A required
preponderantly exchange forces.

Additional evidence for a Serber type force came from

the work of Christian, Hart and Noyes[2] who, in Berkeley,
were working, as the experimental results came in, to find
phenomenological forces which would fit the data. They had
no difficulty in reproducing the shape of the scattering
curve shown in Fig 9.1, but consistently found a total
cross section about 10% too high, this when only even 1
scattering was included. Addition of some odd wave
scattering would only have increased the discrepancy.

About 90% of the total cross section was due to s wave
scattering. This may seem a little surprising, in view of
the considerable angular dependence shown in Fig 9.1, but
if the s phase shifts are large and the others are small,
the angular dependence will come primarily from
interference between s and higher 1 amplitudes, which is
linear in the higher phase shifts, while the total cross
section depends only quadratically on the small phase
shifts. Thus Christian's difficulty was in finding
interactions which gave s wave phase shifts agreeing with
the ones observed near zero energy and falling off
sufficiently rapidly with increasing energy. The correct
explanation was given by Jastrow,[3] who pointed out that the
correct energy dependence of the s wave phase shift could
be obtained if it were supposed that the nuclear
interaction had a repulsive core. The s wave phase shifts
would then, instead of declining to zero with increasing
energy, actually reverse sign at some finite energy. He
found he could fit the observed total cross section at
90 Mev with a core radius of about 0.5 f.

Later experiments at higher energies on both N-P and
P-P scattering confirmed Jastrow's conclusion. The phase

[2] R.S. Christian and E.W. Hart, Phys. Rev. 77, 441 (1950);
R.S. Christian and H.P. Noyes, Phys. Rev. 79, 85 (1950)
[3] R. Jastrow, Phys. Rev. 79, 389 (1950); Phys. Rev. 81, 165
(1950)

shifts deduced from these measurements are shown in Fig 9.2.[*] It will be seen that the $^3S_0$ phase shift changes sign at about 350 Mev and the $^1S_0$ at about 250 Mev.

The argument based on symmetry of the scattering around 90° implicitly assumed central forces, and we should consider the effect of non-central forces in the triplet states. The scattering amplitude can be written as a matrix in an M_L, M_S representation. For an incoming beam in the z direction $M_L=0$, and the scattering amplitude is

$$f_{M_L' M_S'; 0 M_S} = (1/2ip) \Sigma_1 (2l+1) f^1_{M_L' M_S'; 0 M_S} Y_{1 M_L'}(\theta, \emptyset), \qquad (9.01)$$

with

$$Y_{1 M_L'}(\theta, \emptyset) = [(1+M_L')!/(1-M_L')!]^{1/2} P_1^{M_L'}(\cos\theta) e^{i M_L'}.$$

Since $M = M_L + M_S$ is a constant of the motion, $M_L' + M_S' = M_S$.

In the situation we are considering, large s wave phase shift and others small, in the cross section the first order interference term between s wave and higher waves will be proportional to the non-s wave part of $f_{0 M_S; 0 M_S}$. For an unpolarized neutron beam and proton target we observe the average of over M_S,

$$\langle f_{00} \rangle = (1/3) \Sigma_{M_S} f_{0 M_S; 0 M_S}. \qquad (9.02)$$

Now let us suppose that, since the phase shifts are small, we can calculate the non-s part of f in Born approximation. Using plane wave Born approximation, $f_{0 M_S; 0 M_S}$ is proportional to $V_{\underline{p}' M_S; \underline{p} M_S}$. Summing over M_S, as in (9.02), is equivalent to averaging over directions of $\underline{S}$, so for any V whose classical average (i.e. regarding operators as c-numbers) over spin directions is zero, $\Sigma_{M_S} V_{\underline{p}' M_S; \underline{p} M_S} = 0$, and there will be no contribution to $\langle f_{00} \rangle$.

---

[*] Reprinted from A. Bohr and B.R. Mottelson, Nuclear Structure, Benjamin (1969), v.1, p.264.

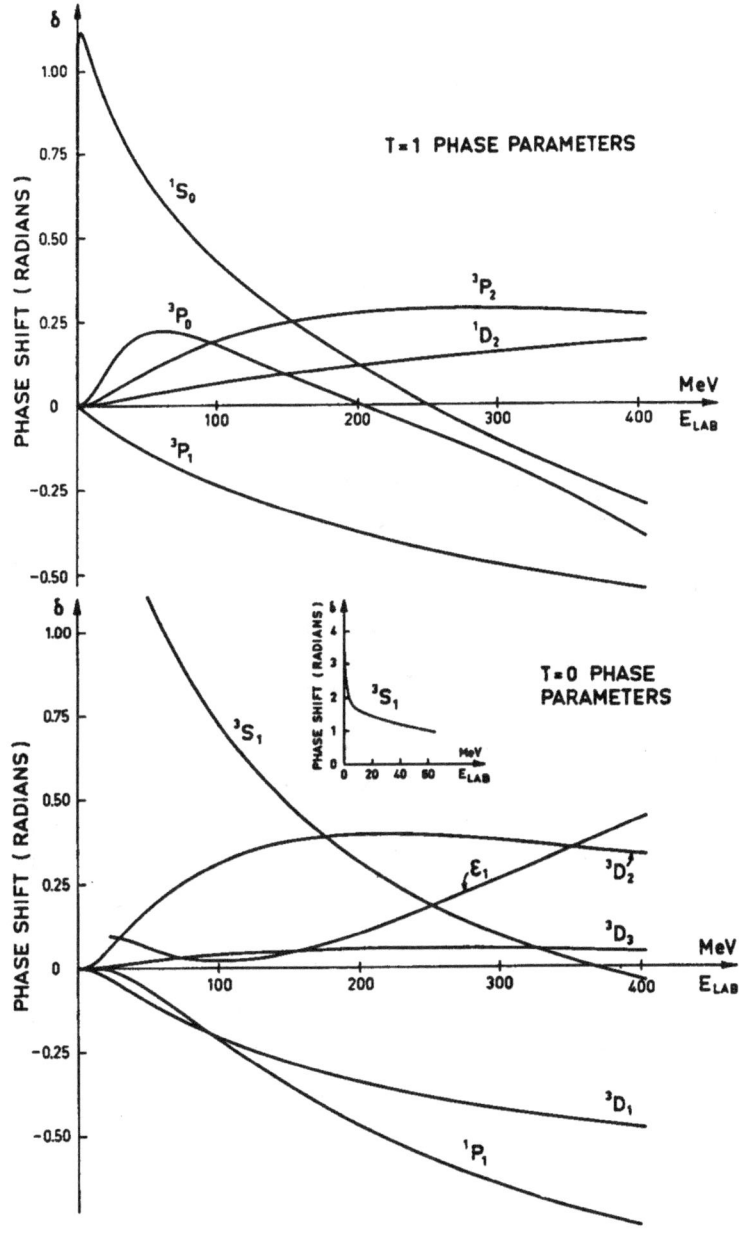

Fig. 9.2

Our previous conclusion that the front back asymmetry of the 90 Mev N-P scattering indicates small odd wave forces was thus an overstatement: what it indicates is small odd wave central forces.

It is instructive to consider in more detail the simplest case of a non-central force, that is, a spin-orbit force. Suppose there is both a central and spin-orbit force, so the potential is of the form

$$V_c(r) + V(r)\underline{S} \cdot \underline{L}.$$

In plane wave Born approximation the spin-orbit term gives a scattering amplitude proportional to

$$i\underline{S} \cdot (\underline{p}' \times \underline{p}),$$

which, since $(\underline{p}' \times \underline{p})$ is perpendicular to $\underline{p}$ (i.e. to $\underline{z}$), involves S_x and S_y but not S_z. There are thus no diagonal matrix elements and the contribution to $\langle t_{00} \rangle$ is zero, as expected.

Looked at in another way: for a central field alone, the exact form of f^1 is

$$f^1 = (e^{2i\delta_1} - 1)\underline{1},$$

where $\underline{1}$ is a 3x3 unit matrix. In the presence of the spin-orbit force the potential is different in each of the three j states, j=1+1,1,1-1. There are three different phase shifts, δ_j^1. We can write the amplitude matrix using the same technique we used in Sec 3II to find the N-P coherent amplitude,

$$f^1 = \Sigma_j (e^{2i\delta_j^1} - 1)P_j,$$

where P_j is the projection operator for the j'th state. The contribution to $\langle f_{00} \rangle$ is proportional to

$$\langle f_{00} \rangle = \Sigma_j (e^{2i\delta_j^1} - 1)\langle (P_j)_{00} \rangle,$$

with

$$\langle (P_j)_{00} \rangle = (1/3)\Sigma_{M_S} (P_j)_{0M_S;0M_S}.$$

Since P_j is invariant under any rotation of both spin and orbit, its average over spin directions is invariant under orbital rotations, and so is independent of M_L. We can therefore write

$$\langle (P_j)_{00} \rangle = \frac{1}{3(2l+1)} \Sigma_{M_L M_S} (P_j)_{M_L M_S; M_L M_S} = \frac{1}{3(2l+1)} \text{Tr}(P_j).$$

The trace is independent of representation; using the j, M_j representation $P_j=1$ for each of the $2j+1$ states having the correct j and $P_j=0$ for the others, so $\text{Tr}(P_j)=2j+1$. Thus

$$\langle f^1_{00} \rangle = \frac{1}{3(2l+1)} \Sigma_j (2j+1)(e^{2i\delta^1_j}-1). \qquad (9.03)$$

Equation (9.03) is exact. Now let us suppose that the phase shifts due to the spin-orbit potential are small, though not necessarily those due to the central potential, so

$$\delta^1_j = \delta^c_l + \delta^{1\prime}_j, \qquad (9.04)$$

where δ^c_l is the phase shift that would be produced by the central force acting alone. Equation (9.03) gives

$$\langle f^1_{00} \rangle = (e^{2i\delta^c_l}-1) + \frac{2i}{3(2l+1)} e^{2i\delta^c_l} \Sigma_j (2j+1)\delta^{1\prime}_j. \qquad (9.05)$$

In Born approximation $\delta^{1\prime}_j$ is proportional to a diagonal matrix element of the non-central potential; the mean matrix element appearing in the second term of (9.05) is, as the argument leading to (9.03) shows, also the mean over M_S and M_L, and equals the classical mean over spin and orbital directions.[a] A non-central force having zero classical average will give no contribution to (9.05). This can readily be verified explicitly for a spin-orbit force; for this

$$\delta^{1\prime}_j = \epsilon(\underline{l} \cdot \underline{S}) = (1/2)\epsilon[j(j+1)-l(l+1)-S(S+1)], \qquad (9.06)$$

which, when substituted in (9.05), gives a zero sum. For a tensor force

$$\delta^{1\prime}_j = \epsilon P_2(\cos\theta_{lS}), \qquad (9.07)$$

with $P_2(\cos\theta_{lS})$ given by (6.09); again the contribution to

---

[a] See J.H. Van Vleck, Theory of Electric and Magnetic Susceptibilities, Oxford (1932), p.137.

(9.05) is zero. For example, for l=1 (9.06) gives $\delta_2^1, \delta_1^1, \delta_0^1$ in the ratio 1:-1:-2, while (9.07) gives 1:-5:10. For l=2, (9.06) gives $\delta_3^2, \delta_2^2, \delta_1^2$ in the ratio 2:-1:-3, while (9.07) gives 2:-7:7. In all four cases

$$\Sigma_j (2j+1)\delta_j^{l'} = 0. \tag{9.08}$$

These intervals are illustrated in Fig 9.3.

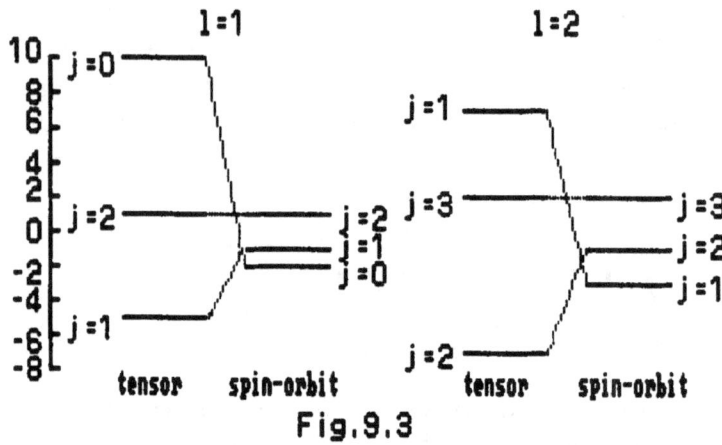

Fig.9.3

The scattering formula for a tensor force is more complicated than for a spin-orbit force because a tensor force mixes states of different l. Thus an incoming s wave will give rise to outgoing s and d waves and vice-versa. However, in calculating $\langle f_{00} \rangle$ one finds the mixing terms also vanish on averaging over M_S.

It follows from (9.04) and (9.08) that the mean phase shift

$$\langle \delta_j^l \rangle = \frac{1}{3(2l+1)} \Sigma_j (2j+1)\delta_j^l = \delta_l^c.$$

The ^3P phase shifts shown in Fig 9.1 have a mean nearly equal to zero for all energies, showing the absence of central forces in the triplet-odd states. The ^3D phase shifts have a positive mean, confirming the attractive central forces in the triplet-even states. We can also get, by inspection, a notion of the nature of the non-central

forces. As we see from Fig 9.3, there is a characteristic
difference between the spin-orbit and the tensor cases. For
spin-orbit the j values are in order, while for tensor the
lowest j value crosses over the other two. The ^{3}P phase
shifts show the tensor ordering at low energy, as one would
expect from the one-pion-exchange potential, (8.09).
However, at energies above 100 Mev one sees a cross over of
the j=0 and j=2 phase shifts, indicating the presence of a
shorter range spin orbit force. Note also that at low
energy the ordering of the ^{3}P states is that shown in
Fig 9.2, while the ^{3}D state ordering is inverted. Such an
inversion is to be expected for the potential (8.09)
because the $\tau_1 \cdot \tau_2$ factor has opposite sign for T=1 and T=0
states.

        A number of potentials have been designed to give the
observed phase shifts (see Bohr and Mottelson, v.1, p. 266,
for a couple of examples). A general description of the
findings is this: at larger distances there is a tensor
force as given by the one-pion-exchange potential, there is
a repulsive core of radius 0.5 f, a short range spin-orbit
force and, in the even l states, an additional strong short
range central force.

SEC.9II HIGHER ENERGY REACTIONS

This section consists of the Appendices B,C and D.

CHAPTER 10

RESONANCE LEVELS

SEC.10I DECAY OF A RESONANCE LEVEL

The problems of the decay of a resonance level and of the scattering of a particle by a resonance level are basic ones in physics. We will illustrate the problems by a simple model, in which we suppose that a system at rest in state A can decay to another in state B with emission of a particle b,

$$A \rightarrow B+b.$$

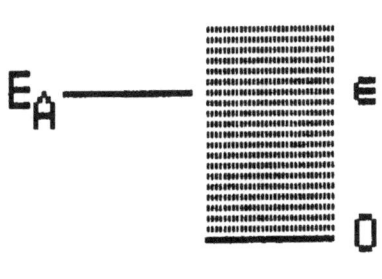

Fig.10.1

The states ϕ_A, with energy E_A, and ϕ_ϵ, which represents B+b with energy ϵ, are supposed to be the eigenfunctions and eigenvalues of a Hamiltonian H_0. The spectrum is as shown in Fig.10.1: for convenience the minimum energy of the continuous spectrum is taken to be $\epsilon=0$, and the continuum

overlaps E_A, i.e. $E_A>0$. The exact Hamiltonian is
$$H=H_0+V,$$
where V has matrix elements between $\emptyset_A$ and $\emptyset_\epsilon$, corresponding to emission of b, and Hermitian conjugate elements corresponding to absorption, i.e. the reaction with the arrow reversed. An example would be the decay of a hydrogen atom from the 2p to the 1s state with emission of a photon. Here $V=-\underline{j}\cdot\underline{A}$, with $\underline{j}$ the electric current and $\underline{A}$ the transverse vector potential. The real electrodynamic problem differs from our model in two respects: V has matrix elements to other states of the hydrogen atom, which are being ignored, and also both the 2p and 1s states can emit and absorb photons, whereas in the model the 2p can only emit and the 1s can only absorb. As a consequence, the exact solution of the real problem would require the inclusion of many-photon states.

The wave function for our model problem can be specified by the probability amplitudes of the states, $\underline{c}(t)=(c_A(t),c_\epsilon(t))$. The states $\emptyset_\epsilon$ are taken to be normalized with respect to energy, so
$$|c_A|^2+\int_0^\infty|c_\epsilon|^2d\epsilon=1.$$
For the decay problem we suppose that at t=0 the system is in state A, i.e. $c_A(0)=1$, $c_\epsilon(0)=0$. The time dependence of the amplitudes is governed by the Schrodinger equation, $i\partial\underline{c}/\partial t=H\underline{c}$, which, written out explicitly, is
$$idc_A/dt=E_Ac_A+\int_0^\infty V_{A\epsilon}c_\epsilon d\epsilon,$$
$$idc_\epsilon/dt=\epsilon c_\epsilon+V_{\epsilon A}c_A. \qquad (10.01)$$

We seek solutions in terms of Fourier integrals,
$$c_A=(1/2\pi i)\int_{-\infty}^\infty c_{AE}e^{-iEt}dE,$$
$$c_\epsilon=(1/2\pi i)\int_{-\infty}^\infty c_{\epsilon E}e^{-iEt}dE. \qquad (10.02)$$

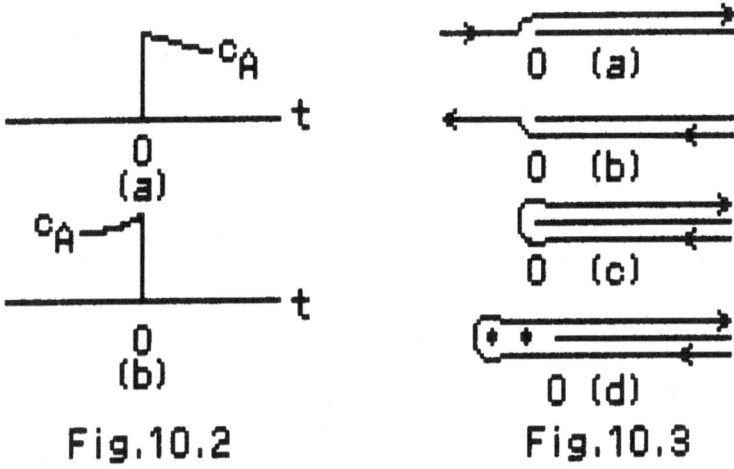

Fig.10.2   Fig.10.3

To find a solution we introduce a mathematical device: we first seek a solution which equals δ for $t>0$, but is zero for $t<0$, so c_A jumps from 0 to 1 at $t=0$ (Fig.10.2a). At the discontinuity $dc_A/dt=\delta(t)$, so the first of equations (10.01) is changed by addition of a term $i\delta(t)$ on the right hand side. The Fourier transforms then satisfy

$$Ec_{AE}=E_A c_{AE}+\int_0^\infty V_{A\epsilon}c_{\epsilon E}d\epsilon-1,$$
$$Ec_{\epsilon E}=\epsilon c_{\epsilon E}+V_{\epsilon A}c_{AE}.$$

The second of these equations gives

$$c_{\epsilon E}=-[V_{\epsilon A}/(\epsilon-E)]c_{AE}, \qquad (10.03)$$

and substituting this back into the first equation gives

$$c_{AE}=1/(E_A-E-\Sigma(E)), \qquad (10.04)$$

where

$$\Sigma(E)=\int_0^\infty |V_{\epsilon A}|^2/(\epsilon-E)d\epsilon. \qquad (10.05)$$

If we consider $\Sigma(E)$ as a function of the complex variable E, we see that, because of the pole in the integrand, it will have a discontinuity as E crosses the positive real axis. If a cut is made on the positive real axis $\Sigma(E)$ is analytic in the cut plane. If we suppose that c_{AE} and $c_{\epsilon E}$ are also analytic in the cut plane the condition $\delta(t)=0,t<0$ will be satisfied provided we take the

E integrations in (10.02) to pass above the cut as shown
in Fig.10.3a, since for negative t the contours can be
closed in the upper half plane and, since the integrands
are analytic, the integrals are zero.

We could similarly obtain a solution which equals $\underline{c}$
for t<0 and is zero for t>0, for which c_A jumps from 1 to 0
at t=0 as in Fig.10.2b, by choosing the contour of
Fig.10.3b. The direction is reversed to take account of the
reversed sign of the δ-function. The sum of the two
solutions gives $\underline{c}$ for all t. Adding the contours (a) and
(b) gives the contour (c) of Fig.10.3, and

$$c_A(t)=(1/2\pi i)\int_c c_{AE}e^{-iEt}dE,$$
$$c_\epsilon(t)=(1/2\pi i)\int_c c_{\epsilon E}e^{-iEt}dE. \qquad (10.06)$$

This is an example of the solution of the Schrodinger
equation for any Hamiltonian, $\underline{H}$, with the boundary
condition that at t=0 $\underline{c}(t)=\underline{c}_0$. The solution of $i\partial\underline{c}/\partial t=\underline{H}\underline{c}$ is
given by the operator equation

$$\underline{c}(t)=e^{-i\underline{H}t}\underline{c}_0, \qquad (10.07)$$

or, following an argument similar to that given above, by

$$\underline{c}(t)=(1/2\pi i)\int_c [e^{-iEt}/(\underline{H}-E)]\underline{c}_0 dE. \qquad (10.08)$$

Since the eigenvalues of $\underline{H}$ are real, the singularities of
the integrand are on the real E axis, consisting of poles
at the points where $\underline{H}$ has discrete eigenvalues and cuts
where there is a continuous spectrum. The contour c
encloses both poles and cuts, as in Fig.10.3d.

The supposition made above that c_{AE} and $c_{\epsilon E}$ are
analytic in the the cut plane thus follows from the
Hermiticity of H, with the modification that a pole exists
on the negative real axis if H has a bound state.
Examining (10.05) we see that $\Sigma(E)$ is real and positive on
the negative real axis, and is monotonically decreasing as

$E \to -\infty$. It follows that c_{AE}, given by (10.04), will have one pole on the negative real axis if $E_A - \Sigma(0) < 0$, in which case there is a discrete bound state. If $E_A > \Sigma(0)$ there is no bound state; in discussing the decay problem we shall suppose this is the case, i.e. that V is not large enough to produce binding. For the decay problem the solutions (10.03) and (10.04) may be regained from (10.08) by expanding $[1/(\underline{H}-E)]\underline{\delta}_0 = [1/(\underline{H}_0 + \underline{V}-E)]\underline{\delta}_A$ in powers of $\underline{V}$ and summing the resultant geometric series.

If $\underline{\delta}_0$ is expanded in terms of the stationary state wave functions of $\underline{H}$, which we suppose has discrete eigenvalues at negative energies E_i and a continuum for $E>0$,

$$\underline{\delta}_0 = \Sigma_i b_{E_i} \underline{\delta}_{E_i} + \int_0^\infty b_E \underline{\delta}_E dE,$$

it follows from either (10.07) or (10.08) that

$$\underline{\delta}(t) = \Sigma_i b_{E_i} \underline{\delta}_{E_i} e^{-iE_i t} + \int_0^\infty b_E \underline{\delta}_E e^{-iEt} dE. \qquad (10.09)$$

We mention this well known result in order to bring out the converse, that if (10.08) (or (10.06)) is written in the form (10.09), as a sum of residues plus an integral from 0 to ∞, the coefficients of $\exp(-iE_i t)$ and $\exp(-iEt)$ give the stationary state wave functions of $\underline{H}$, aside from a factor b_E. Thus the solution of (10.08), for any $\underline{\delta}_0$, also determines for us all the stationary states for which b_E is not zero. (However, if there is degeneracy, only a particular linear combination of the degenerate states is determined.)

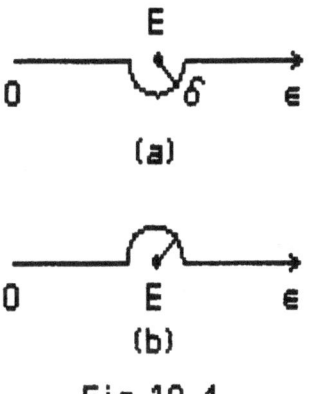

(a)

(b)

Fig.10.4

We turn now to a discussion of the function $\Sigma(E)$, defined by (10.05). As E approaches the positive real axis from above the integration path can be deformed, as shown in Fig.10.4a, into an integral from 0 to E-δ, plus an integral from E+δ to ∞, plus the integral around the semicircle of radius δ. In the limit $\delta\to0$ the sum of the first two gives the principle value of the integral while the third gives πi times the residue at E. Thus, on the positive real axis

$$\Sigma(E+i0)=P\int_0^\infty |V_{\epsilon A}|^2/(\epsilon-E)d\epsilon+i\pi|V_{EA}|^2.$$

As E approaches the real axis from below the ϵ contour is taken as in Fig.10.4b, and

$$\Sigma(E-i0)=P\int_0^\infty |V_{\epsilon A}|^2/(\epsilon-E)d\epsilon-i\pi|V_{EA}|^2.$$

Let us define the function $\gamma(E)$ by analytic continuation of

$$\gamma(E)=\pi|V_{EA}|^2,$$

and $\Delta(E)$ by

$$\Sigma(E)=\Delta(E)\pm i\gamma(E), \qquad (10.10)$$

the plus sign being taken if E is in the upper half plane, the minus in the lower half plane. Note that $\gamma(E)$ and $\Delta(E)$ are continuous as E crosses the real axis; cuts that may be required to make these functions single valued will be taken on the negative real axis.

An example may be helpful. If

$$|V_{\epsilon A}|^2=\alpha\epsilon^{1/2}/(\epsilon+\mu)^2$$

the integral in (10.05) is elementary, and one finds

$$\gamma(E)=\pi\alpha E^{1/2}/(E+\mu)^2, \quad -\pi<\arg E<\pi,$$
$$\Delta(E)=(\pi\alpha/\mu^{1/2})[(\mu/(E+\mu)^2)-1/(2(E+\mu))].$$

It should be noted that,despite appearances, Σ has no pole at E=$-\mu$. It is easily verified that $\Sigma(-\mu)=\pi\alpha/(8\mu^{3/2})$.

The equations (10.06) can also be written, using (10.04), (10.03) and (10.10)

$$c_A=\frac{1}{2\pi i}\int_0^\infty [\frac{1}{E_A-E-\Delta(E)-i\gamma(E)} - \frac{1}{E_A-E-\Delta(E)+i\gamma(E)}]e^{-iEt}dE,$$

$$c_\epsilon=-\frac{1}{2\pi i}\int_0^\infty V_{\epsilon A}[\frac{1}{(\epsilon-E-i0)(E_A-E-\Delta(E)-i\gamma(E))}$$
$$-\frac{1}{(\epsilon-E+i0)(E_A-E-\Delta(E)+i\gamma(E))}]e^{-iEt}dE. \qquad (10.11)$$

In either case the first term in the integrand comes from
the contour along the upper side of the cut, the second
from the lower. The $-i0$ in the first term for c_ϵ indicates
that the contour of integration passes above the point ϵ,
the $+i0$ in the second that the contour passes below ϵ. Such
a specification is not required for the other factor in the
denominator because of the imaginary $i\gamma(E)$ term. If $\gamma(E)$ is
small the other factor in the first term will give a pole
in the fourth quadrant near the real axis. This does not
contradict the analyticity of Σ, c_{AE} and $c_{\epsilon E}$ in the cut
plane because the the pole is reached by crossing the cut
from above and analytically continuing the functions into
the fourth quadrant. The pole thus lies in another sheet of
the function. Similarly, the second term in the integrand
will have a pole in the first quadrant reached by crossing
the cut from below.

Let the position of of the pole in the fourth quadrant
be

$$E_P=E_A^r-i\Gamma/2. \qquad (10.12)$$

The energy E_A^r is the perturbed energy of the state A and
Γ, as we shall shortly see is the decay rate of the state.
The condition determining E_P is

$$E_A-E_P-\Delta(E_P)-i\gamma(E_P)=0.$$

Let $E_P=E_A^r+e$ and expand $\Delta(E)$ and $\gamma(E)$ in Taylor series in
e. To first order in e,

$$E_A-E_A^r-e-\Delta(E_A^r)-\Delta^r(E_A^r)e-i\gamma(E_A^r)-i\gamma^r(E_A^r)e=0, \qquad (10.13)$$

where $\Delta^r(E_A^r)=d\Delta(E_A^r)/dE$ and $\gamma^r(E_A^r)=d\gamma(E_A^r)/dE$. Solving
for e gives

$$e=N(E_A^r)(s-i\gamma(E_A^r)), \qquad (10.14)$$

with

$$s=E_A-E_A^r-\Delta(E_A^r)$$

and

$$N(E)=1/(1+\Delta^r(E)+i\gamma^r(E)). \qquad (10.15)$$

It may be noted that we can also write N(E) in a form which
will be recognized as related to a normalization factor,

$$N(E)=1/(1+\Sigma'(E))=1/[1+\int_0^\infty |V_{\epsilon A}|^2/(\epsilon-E)^2 d\epsilon],$$

with, in (10.14), $N(E_A')=N(E_A'+i0)$.

Rationalizing $N(E_A')$, we have

$$e=(1+\Delta'(E_A')-i\Upsilon'(E_A'))(s-i\Upsilon(E_A'))/[(1+\Delta'(E_A'))^2+\Upsilon'(E_A')^2]$$

$$=\frac{(1+\Delta'(E_A'))s-\Upsilon'(E_A')\Upsilon(E_A')-i[(1+\Delta'(E_A'))\Upsilon(E_A')+\Upsilon'(E_A')s]}{(1+\Delta'(E_A'))^2+\Upsilon'(E_A')^2}.$$

According to (10.12) e should have no real part, so

$$s=\Upsilon'(E_A')\Upsilon(E_A')/(1+\Delta'(E_A')),$$

which, from the definition of s following (10.14), gives

$$E_A'=E_A-\Delta(E_A')-\Upsilon'(E_A')\Upsilon(E_A')/(1+\Delta'(E_A')).$$

With this value of s, e becomes

$$e=-i\Upsilon(E_A')/(1+\Delta'(E_A')),$$

so

$$\Gamma=2\Upsilon(E_A')/(1+\Delta'(E_A'))=2\Upsilon(E_A')/[1+\mathrm{Re}\int_0^\infty |V_{\epsilon A}|^2/(\epsilon-E_A')^2 d\epsilon].$$

$$(10.16)$$

Thus the lowest order perturbation theory value, $\Gamma=2\Upsilon(E_A)$, is modified in two respects: first, Υ is evaluated at the perturbed value, E_A', rather than at E_A; second, there is the renormalizing factor exhibited in (10.16).

The validity of the Taylor series expansion made in (10.13) is suspect if $|\Delta'|$ and $|\Upsilon'|$ are not small compared to unity. To estimate these quantities we must consider the properties of $|V_{\epsilon A}|^2$. First we observe that the convergence of the integral in formula (10.05) for $\Sigma(E)$ requires that $|V_{\epsilon A}|^2\to 0$ as $\epsilon\to\infty$, so there must be an energy, μ, beyond which $|V_{\epsilon A}|^2$ starts to decrease. The behavior of $|V_{\epsilon A}|^2$ as $\epsilon\to 0$ is also significant. This behavior depends on the problem; as an example let us consider nuclear emission of a neutron (an uncharged non-relativistic particle) in an s state. The perturbation theory formula for the transition rate is perhaps more familiar in the form "2π times matrix element squared times density of final states", i.e. $\Gamma=2\pi|V_{pA}|^2\cdot 4\pi p^2/v$, where the normalization of the final states is to p, rather than ϵ. For the s wave case we see, from $|\emptyset_\epsilon|^2 d\epsilon = |\emptyset_p|^2 \cdot 4\pi p^2 dp$, that $\emptyset_\epsilon = (4\pi p^2/(d\epsilon/dp))^{1/2}\emptyset_p$, hence

$$V_{\epsilon A} = (4\pi p^2/v)^{1/2} V_{pA}. \qquad (10.20)$$

With energy normalization, the density of states factor is included in $|V_{\epsilon A}|^2$. Moreover, since $|V_{pA}|^2$ approaches a constant as $p \to 0$, $|V_{\epsilon A}|^2$ is proportional to p, or $\sqrt{\epsilon}$, as $\epsilon \to 0$. If we were considering decay into a state of angular momentum l, we would expect an additional factor $(pR)^{2l}$, with R the radius of the decaying system, in $|V_{\epsilon A}|^2$, and thus

$$\gamma(\epsilon) = k\epsilon^{l+1/2}. \qquad (10.18)$$

In the case $E_A' < \mu$ and s state decay the relative magnitudes of the various quantities can easily be written down:

$$\gamma(E_A') \approx k\sqrt{E_A'},$$

$$\Delta(E_A') \approx k\sqrt{\mu} \approx \gamma(E_A')\sqrt{(\mu/E_A')},$$

$$\gamma'(E_A') \approx \gamma(E_A')/2E_A',$$

$$|\Delta'(E_A')| \approx \Delta(E_A')/\mu \approx \gamma'(E_A')\sqrt{(E_A'/\mu)}.$$

The conditions for validity of the calculation leading to (10.16), that $\gamma'(E_A') \ll 1$ and $\Delta'(E_A') \ll 1$, thus depend on $\gamma(E_A')/E_A'$ being small (lifetime long compared to the period $1/E_A'$).

Our purpose now is to evaluate $c_A(t)$ for large t, by which we mean $E_A't \gg 1$. This is accomplished by deforming the path of integration in (10.11) to the negative imaginary axis, so the limits of integration become 0 to $-i\infty$. Putting $E=-iy$, we obtain

$$c_A(t) = N(E_p)e^{-iE_p t}$$
$$- \frac{1}{\pi}\int_0^\infty \frac{\gamma(-iy)e^{-yt}}{(E_A-\Delta(-iy)+iy)^2 + \gamma(-iy)^2} dy + O(e^{-\mu' t}). \qquad (10.19)$$

The first term on the right comes from the residue of the

pole at E_P in the second sheet. The factor $N(E_P)$, which arises in evaluating the residue, is given by

$$N(E)=1/[(d/dE)(E-E_A+\Delta(E)+i\gamma(E))]=1/(1+\Delta'(E)+i\gamma'(E)),$$

which is the same as (10.15)). We may note that, since derivatives of $\Delta(E)$ and $\gamma(E)$ higher than the first have been neglected in the expansion (10.13), there is little point in our distinguishing between $N(E_P)$ and $N(E_A')$, since their difference is proportional to $\gamma(E_A')$ times second derivatives. A term of order $\exp(-\mu't)$, with $\mu'\approx\mu$, has been indicated to allow for the contingency that the first term of the integrand of (10.11) might have other singularities in the fourth quadrant of the second sheet. For example, if $\gamma(E)$ had a pole in the fourth quadrant the integrand would have a pole close by, say at a point E'. Then $\mu'=-ImE'$, and since the scale of the variation of γ with energy is set by μ, $\mu'\approx\mu$.

Because of the exponential factor in the integrand of (10.19), for large t the contributions to the integral come from $y<1/t$, and an expansion of the rest of the integrand in powers of y is equivalent to an expansion of the integral in powers of $1/(E_A't)$. The leading term is, for $\gamma(E)=k\sqrt{E}$,

$$-\frac{ik(-i)^{1/2}}{\pi(E_A-\Delta(0))^2}\int_0^\infty y^{1/2}e^{-yt}dy=\frac{k}{2\pi^{1/2}(E_A-\Delta(0))^2(it)^{3/2}}. \qquad (10.20)$$

This can also be written

$$\gamma(1/t)/[2\pi^{1/2}(E_A-\Delta(0))^2(i)^{3/2}t] \qquad (10.21)$$

For states of higher l we see, using (10.18), that (10.20) would be proportional to $1/t^{l+3/2}$. Equation (10.21) would still be correct if multiplied by a factor $(-i)^l\Gamma(l+3/2)/\Gamma(3/2)$.

Thus the asymptotic form of $c_A(t)$, for $E_A't\gg1$, is

$$c_A(t)\approx N(E_P)\exp(-iE_Pt)+\gamma(1/t)/[2\pi^{1/2}(E_A-\Delta(0))^2(i)^{3/2}t]$$
$$+O(\exp(-\mu't)). \qquad (10.22)$$

Even for $t=1/E_A'$ the magnitude of the second term on the

right in (10.22) is smaller than the first by a factor $\gamma(E_A')/E_A'$, and for a while it decreases more rapidly with time, e.g. when $t=1/\gamma(E_A')$ it is smaller by a factor $(\gamma(E_A')/E_A')^{5/2}$. Thus, after a transition period of order $1/E_A'$, the probability of the system being in state A is well represented by

$$P_A(t)=|c_A^2(t)| \approx |N(E_P)|^2 e^{-\Gamma t});$$

an exponential decay with an amplitude $|N(E_P)|^2$ and a decay rate $-2ImE_P=\Gamma$. That the amplitude differs slightly from unity is because of transition effects while $t<1/E_A'$.

Of course, if one waits a sufficiently long time the second term on the right in (10.22) will dominate the first, and instead of an exponential decay we will have

$$P_A(t)=const/t^3.$$

If $\gamma(E_A')/E_A'=1/10$ the magnitude of the two terms in (10.22) become equal at twenty one mean lives ($\Gamma t=21$); if $\gamma(E_A')/E_A'=1/100$, at thirty four mean lives.

The ultimate $1/t^3$ behavior is not a peculiarity of our particular problem; rather it is a general feature of the time development of any wave packet. At this point one should recall (10.09): our $\xi(t)$ is a packet of the exact stationary state solutions of the system (in our case there are no bound states; (10.09) has only the integral over the continuous spectrum). This can be illustrated by the simplest case: that of a plane wave packet representing a free particle. If the wave function at time zero is $\xi(r,0)$, the solution of the time dependent Schrodinger equation is[1]

$$\xi(r,t)=(M/2\pi it)^{3/2}\int e^{iM|r-r'|^2/2t}\xi(r',0)dr'. \qquad (10.23)$$

For fixed r, as $t\to\infty$,

$$\xi(r,t)\approx(M/2\pi it)^{3/2}\int \xi(r',0)dr', \qquad (10.24)$$

and we have the typical $1/t^{3/2}$ dependence. The clue to the

---

[1] See, for example, Leonard I. Schiff, Quantum Mechanics, second edition, McGraw Hill, New York, 1955, page 59.

physical reason for this behavior is found in the
observation that the probability amplitude for the particle
having the momentum $\underline{p}$ is

$$c_{\underline{p}}=(2\pi)^{-3/2}\int e^{-i\underline{p}\cdot\underline{r}}\, \xi(\underline{r}',0)d\underline{r}',$$

and thus (10.24) is proportional to the probability
amplitude, $c_{\underline{0}}$, that the particle have zero momentum,

$$\xi(\underline{r},t)\approx(M/it)^{3/2}c_{\underline{0}}. \qquad (10.25)$$

The zero momentum states do not move off in the course of
time and give a constant contribution to the amplitude of
the wave function. We say "zero momentum states" because,
according to the uncertainty relation, in a time t an
energy less than $E=1/t$ cannot be distinguished from zero,
nor can a momentum less than $p=(2ME)^{1/2}=(2M/t)^{1/2}$. The
number of such momentum states is proportional to p^3 or
$(2M/t)^{3/2}$, which accounts for the $(M/t)^{3/2}$ factor in
(10.25). The statement that these states give a constant
contribution to the amplitude depends on pr being less than
one, or (squaring), $2Mr^2/t<1$, which is essentially the
condition for the passage from (10.23) to (10.24).

Another way to make the argument is to observe that
particles starting at the center of a sphere of radius r
will not escape in time t if $vt<r$, or $(p/M)t<r$, or $p<Mr/t$.
The number of particles within the sphere is thus

$$|c_{\underline{0}}|^2\cdot(4\pi/3)p_{max}^3=|c|_{\underline{0}}^2\cdot(4\pi/3)(Mr/t)^3,$$

and their density is this divided by the volume of the
sphere, $(4\pi/3)r^3$,

$$\text{density}=(M/t)^3|c_{\underline{0}}|^2, \qquad (10.26)$$

which is just $|\xi|^2$, as given by (10.25).

The second term in (10.22) can be understood in the
same way. Within any fixed radius around the point of
decay, after a sufficiently long time the contributions to
the integral in (10.09) come, because of the rapid
oscillation of the factor $\exp(-iEt)$, from $E<1/t$, and $\xi(t)$
is of order of magnitude

$$\bar{s}(t) \approx b_E \bar{s}_E \delta E,$$

with $E = \delta E = 1/t$. Simple first order perturbation theory gives $b_E = V_{EA}/(E - E_A)$, or for $E \approx 0$,

$$b_E \approx -V_{EA}/E_A.$$

Also, according to perturbation theory, the exact stationary state, $\bar{s}_E$, contains $\emptyset_A$ with an amplitude

$$-V_{AE}/(E_A - E) \approx -V_{AE}/E_A$$

for $E \approx 0$. The amplitude of $\emptyset_A$ in the wave packet is thus

$$c_A(t) \approx (|V_{EA}|^2/E_A^2) \delta E = \gamma(1/t)/(\pi E_A^2 t),$$

agreeing in all significant respects, with (10.22). If we consider proton, rather than neutron, emission by a nucleus, the above result is altered because, for small E, $\gamma(E)$ is reduced by the Coulomb barrier penetration factor. According to the argument following (7.18) we shall have, in place of $\gamma(E) = kE^{1/2}$,

$$\gamma(E) = kE^{1/2} \cdot 2\pi\eta(E)/(e^{2\pi\eta(E)} - 1),$$

with $\eta(E) = Z_B Z_b e^2/v = Z_B Z_b e^2 (M/2E)^{1/2}$. The integral that then appears in (10.20) can be done by the saddle point method and we find that (10.21) is replaced by

$$\frac{(-i)^{5/6} k\eta(1/2t)^{4/3}}{(3\pi)^{1/2}(E_A - \Delta(0))^2 t^{3/2}} e^{[-(3/2)i^{1/3}\eta(1/2t)^{2/3}]}$$

In place of $P_A \approx 1/t^3$, we shall have

$$P_A \approx t^{-5/3} e^{-at^{1/3}}.$$

Other considerations can also influence the result. For example, if particle B is not free to fall under gravity, while b is, the gravitational field will sweep away the zero energy particles. In this case the analytic behavior in (10.08) is quite different, since, with a constant gravitational field, the spectrum of the eigenvalues of H extends from $E = -\infty$ to $E = \infty$, rather than rom $E = 0$ to $E = \infty$.

Returning to the equations (10.11): $c_\epsilon(t)$ can be evaluated for $t \gg 1/E_A'$ just as was $c_E(t)$. The first term in the integrand now has two poles, so in place of the first line of (10.19) we have

$$c_\epsilon(t) \approx [V_{\epsilon A}/(\epsilon - E_A + \Delta(\epsilon) + i\gamma(\epsilon))]e^{-i\epsilon t}$$
$$-[N(E_P)V_{E_P A}/(\epsilon - E_P)]e^{-iE_P t}.$$

The energy distribution of the emitted particles is

$$P_\epsilon d\epsilon = |c_\epsilon(\infty)|^2 d\epsilon = \gamma(\epsilon)/\{\pi[(\epsilon - E_A + \Delta(\epsilon))^2 + \gamma(\epsilon)^2]\}d\epsilon. \qquad (10.27)$$

It is not difficult to verify that the total probability is unity. We can write

$$\int_0^\infty P_\epsilon d\epsilon = -(1/2\pi i)\int_0^\infty [\frac{1}{\epsilon - E_A + \Delta(\epsilon) + i\gamma(\epsilon)} - \frac{1}{\epsilon - E_A + \Delta(\epsilon) - i\gamma(\epsilon)}]d\epsilon$$

It follows from (10.10) that this can also be written

$$\int_0^\infty P_\epsilon d\epsilon = -(1/2\pi i)\int_C d\epsilon/(\epsilon - E_A + \Sigma(\epsilon)),$$

with the contour that of Fig 10.3c. From (10.04) and (10.06)

$$\int_0^\infty P_\epsilon d\epsilon = c_A(0) = 1.$$

The same result can be obtained by an explicit calculation.

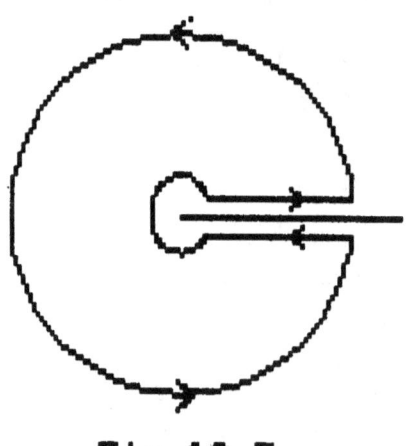

From the analyticity of the integrand we have

$$(2\pi i)\int_{c'} d\epsilon/(\epsilon - E_A + \Sigma(\epsilon)) = 0,$$

where c' is the contour shown in Fig.10.5. Letting the radius of the circle, R, approach infinity, we have

$$(1/2\pi i)\int_C d\epsilon/(\epsilon - E_A + \Sigma(\epsilon))$$

$$= -\lim_{R\to\infty}(2\pi i)\int_R d\epsilon/(\epsilon - E_A + \Sigma(\epsilon))$$

$$= -(2\pi i)\int_R d\epsilon/(\epsilon = -1.$$

Fig.10.5

SEC.10II THE STATIONARY STATES

According to (10.09), the coefficient of exp(-iEt) in (10.11) gives us $b_E \breve{z}_E$, i.e., a multiple of a normalized stationary state, $\breve{z}_E$. To interpret the result let us write it in a coordinate representation. For our example of a nonrelativistic s state system the set of states $\emptyset_\epsilon$, for which the c_ϵ are the probability amplitudes, are of the form

$$\emptyset_\epsilon(r) = (1/4\pi)^{1/2} \emptyset_B \emptyset_b u_\epsilon(r)/r.$$

Here $\emptyset_B$ and $\emptyset_b$ are functions of the internal coordinates of B and b. These internal coordinates can include spin, though we suppose, for the present, that spin and orbital motions are uncoupled. The radial function, $u_\epsilon(r)$, has the asymptotic form for large r

$$u_\epsilon(r) \approx (2/\pi v(\epsilon))^{1/2} \sin(p(\epsilon)r + \delta_P(\epsilon))$$
$$= (1/i)(1/2\pi v(\epsilon))^{1/2} [e^{i(p(\epsilon)r + \delta_P(\epsilon))} - e^{-i(p(\epsilon)r + \delta_P(\epsilon))}].$$
$$(10.28)$$

The factor $(2/\pi v(\epsilon))^{1/2}$, with $v(\epsilon)$ the velocity, $v = d\epsilon/dp = p/M$, normalizes the solutions with respect to ϵ, i.e., so

$$\int_0^\infty u_{\epsilon'}(r)^* u_\epsilon(r) dr = \delta(\epsilon' - \epsilon).$$

The phase shift $\delta_P(\epsilon)$ is that caused by whatever potential acts between B and b, other than the interaction with the resonant state. In the coordinate representation the wave function is

$$\breve{z}(r,t) = \int c_\epsilon \emptyset_\epsilon(r) d\epsilon + c_A(t) \emptyset_A.$$ $$(10.29)$$

If we insert the values of $c_\epsilon(t)$ and $c_A(t)$ given by (10.11) and pick out the coefficient of exp(-iEt) we obtain

$$b_E \breve{z}_E(r) =$$
$$- \frac{1}{2\pi i} \int V_{\epsilon A} \left[\frac{1}{(\epsilon-E-i0)(E_A-E-\Delta(E)-i\gamma(E))} \right.$$
$$\left. - \frac{1}{(\epsilon-E+i0)(E_A-E-\Delta(E)+i\gamma(E))} \right] \emptyset_\epsilon d\epsilon$$
$$+ \frac{\gamma(E)}{\pi[(E_A-E-\Delta(E))^2 + \gamma(E)^2]} \emptyset_A.$$ $$(10.30)$$

To evaluate b_E we look at the asymptotic form of the wave function for large r,

$$b_E \delta_E(r) \approx - \frac{\phi_B \phi_b}{(4\pi)^{1/2} 2\pi i r} \int_0^\infty \frac{v(\epsilon) dp \ V_{\epsilon A}}{i(2\pi v(\epsilon))^{1/2}} [\ldots]$$
$$\cdot [e^{i(p(\epsilon)r + \delta_P(\epsilon))} - e^{-i(p(\epsilon)r + \delta_P(\epsilon))}]. \qquad (10.31)$$

Here [...] is the bracketed factor in (10.30). Note that the integration variable has been changed from ϵ to p, $d\epsilon = v(\epsilon) dp$. We can now use the method we used to evaluate $c_A(t)$ for large t to evaluate $b_E \delta_E(r)$ for large r. For the term proportional to exp(ipr) the contour can be deformed to the positive imaginary axis. The only contribution to the leading term in the asymptotic expansion, that proportional to 1/r, comes from the pole on the real axis in the first term of the square bracket (note that the contour passes below this pole, above the pole in the second term).Similarly, the exp(-ipr) term can be deformed to the negative imaginary axis and the leading contribution comes from the pole on the real axis in the second term of the square bracket. The residues at these poles give

$$b_E \delta_E(r) \approx - \frac{\phi_B \phi_b V_{EA}}{(4\pi)^{1/2} i r (2\pi v(E))^{1/2}}$$
$$\cdot [\frac{e^{i(pr + \delta_P(E))}}{E_A - E - \Delta(E) - i\gamma(E)} - \frac{e^{-i(pr + \delta_P(E))}}{E_A - E - \Delta(E) + i\gamma(E)}]. \qquad (10.32)$$

The condition that δ_E be normalized to E determines the magnitude of b_E but not its phase. We shall chose the phase to satisfy the outgoing wave scattering condition (incoming plane wave plus outgoing spherical wave), i.e., we chose the phase so the incoming spherical wave has no phase shift. We then have

$$b_E = -V_{EA} e^{-i\delta_P(E)} / (E_A - E - \Delta(E) + i\gamma(E))$$

and

$$\delta_E \approx \frac{\phi_B \phi_b}{(4\pi)^{1/2} i r (2\pi v(E))^{1/2}} [e^{i(p(E)r + 2\delta_R(E) + 2\delta_P(E))} - e^{-ip(E)r}],$$

where

$$e^{2i\delta_R(E)} = (E_A - E - \Delta(E) + i\gamma(E))/(E_A - E - \Delta(E) - i\gamma(E)), \qquad (10.33)$$

or, equivalently,

$$\tan\delta_R(E) = \gamma(E)/(E_A - E - \Delta(E)). \qquad (10.34)$$

The energy distribution of the wave packet, which is also that of the emitted particles, is given by

$$P(E)dE = |b_E|^2 dE = \gamma(E)/\{\pi[(E_A - E - \Delta(E))^2 + \gamma(E)^2]\}dE. \qquad (10.35)$$

This agrees with our previous result, (10.27). The phase shift of the outgoing wave is $\delta(E) = \delta_R(E) + \delta_P(E)$, from which the scattering cross section of particle b by the system in state B can be determined. Before discussing the scattering, however, we shall continue our study of the decay problem by examining the behavior of $\xi(r,t)$ at large distances from the source.

SEC.10.III THE DECAY WAVE FUNCTION AT LARGE r

The wavefunction $\bar{Q}(r,t)=\int_0^\infty b_E \bar{Q}_E dE$ has the asymptotic form, at large r,

$$\bar{Q}(r,t)\approx -\frac{\emptyset_B\emptyset_b}{(4\pi)^{1/2}ir}\int_0^\infty \frac{v_{EA}}{(2\pi v(E))^{1/2}}$$
$$\cdot[\frac{e^{i(pr+\delta_P(E))}}{E_A-E-\Delta(E)-i\gamma(E)} - \frac{e^{-i(pr+\delta_P(E))}}{E_A-E-\Delta(E)+i\gamma(E)}]e^{-iEt}dE, \qquad (10.36)$$

as follows from (10.32). A very simple approximation for (10.36) is obtained if we suppose that $\gamma(E_A')/E_A'$ is very small and that the contributions to the integral come from the immediate neighborhood of E_A'. If we let $E=E_A'+\nu$ we can then write

$$\bar{Q}(r,t)\approx\frac{\emptyset_B\emptyset_b v_{E_A'A}}{(4\pi)^{1/2}ir(2\pi v(E_A'))^{1/2}}$$
$$\cdot[e^{i(p(E_A')r-E_A't+\delta_P(E_A'))}\int_{-\infty}^\infty \frac{e^{-i\nu[t-(r/v(E_A'))-\delta_P'(E_A')]}}{\nu+i\gamma(E_A')}d\nu$$
$$-e^{-i(p(E_A')r+E_A't+\delta_P(E_A'))}\int_{-\infty}^\infty \frac{e^{-i\nu[t+(r/v(E_A'))+\delta_P'(E_A')]}}{\nu-i\gamma(E_A')}d\nu],$$

$$(10.37)$$

where $\delta_P'(E_A')=d\delta_P(E_A')/dE$ and, in the exponential factors, we have used

$$p(E)=p(E_A')+(dp(E_A')/dE)\nu=p(E_A')+\nu/v(E_A'). \qquad (10.38)$$

Consider the first integral in (10.37). If $t-(r/v(E_A'))-\delta_P'(E_A')$ is positive, the contour can be closed in the lower half plane and we obtain the residue at the pole. If it is negative, the contour can be closed in the upper half plane, where the integrand is analytic, and so the result is zero. In the second integral $t+(r/v(E_A'))+\delta_P'(E_A')$ is always positive for $t>0$ and large r, so the contour can always be closed in the lower half plane, where the integrand is analytic, and so the integral is zero. The result is therefore

$$\underline{\xi}(r,t)=0, \quad t<(r/v(E_A'))+\delta_P'(E_A')$$

$$\underline{\xi}(r,t)\approx-\frac{\phi_B\phi_b(2\pi)^{1/2}v_{E_A'A}}{(4\pi)^{1/2}rv(E_A')^{1/2}}e^{i(p(E_A')r-E_A't+\delta_P(E_A'))}$$

$$\cdot e^{-\gamma(E_A')[t-(r/v(E_A'))-\delta_P'(E_A')]}, \quad t<(r/v(E'))+\delta'(E') \quad (10.39)$$

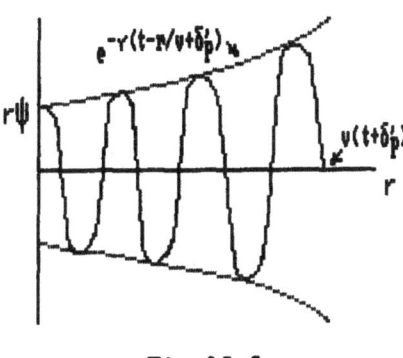

Fig.10.6

Fig.10.6 shows, somewhat schematically, the behavior of $\underline{\xi}(r,t)$. The current of particles through the surface of a sphere of radius r is given by

$$J=4\pi r^2 v(E_A')\int dq_{int}|\underline{\xi}|^2$$
$$=0, \quad t<(r/v(E_A'))+\delta_P'(E_A'),$$
$$=\Gamma e^{-\Gamma[t-(r/v(E_A'))-\delta_P'(E_A')]},$$
$$\qquad t>(r/v(E_A'))+\delta_P'(E_A'),$$

$$(10.40)$$

where $\int dq_{int}$ indicates a summation over the internal coordinates of B and b. Suppose, for the moment, that $\delta_P'(E_A')=0$. Eq.(10.40) then tells us that there is no current before the time, $r/v(E_A')$, that it takes a particle of velocity $v(E_A')$ to reach r. Thereafter the current falls exponentially with a mean life $1/\Gamma=1/(2\gamma(E_A'))$.

The $\delta_P'(E_A')$ term is a correction to the arrival time due to the interaction between B and b which gave rise to the phase shift δ_P. This is easily seen if we use the WKB approximation for the phase shift. For a wave of angular momentum l this is given by

$$\delta_P(E)=\lim_{r\to\infty}[\int_{r_t}^r p_V(r)dr-\int_{r_t}^r p_0(r)dr],$$

where

$$p_V(r)=[2ME-2MV(r)-(1+1/2)^2/r^2]^{1/2},$$
$$p_0(r)=[2ME-(1+1/2)^2/r^2]^{1/2},$$

$V(r)$ is the interaction potential and the lower limit in each integral is the turning point. Differentiating with respect to E gives

$$\delta_P'(E) = \lim_{r \to \infty}[\int_{r_t}^{r}(dp_V(r)/dE)dr - \int_{r_t}^{r}(dp_0(r)/dE)dr]$$

$$= \lim_{r \to \infty}[\int_{r_t}^{r}dr/v_V(r) - \int_{r_t}^{r}dr/v_0(r)].$$

Thus $\delta_P'(E)$ is the change in arrival time at large r due to the potential $V(r)$. The effect takes a simpler form for s waves when p is small enough so that the scattering length approximation can be used. We then have $\delta_P = -pa$ and in the exponentials of (10.37) the effect of δ_P' is simply to replace r by r−a.

It is evident that the abrupt rise at $t = r/v(E_A')$ shown in Fig.10.6 and Eq.(10.39) cannot be right if $r/v(E_A')$ is too large. The linear relation between frequency and wavenumber used in (10.38) neglects dispersion, the fact that different portions of the packet move with different group velocities. The packet has an energy spread $\delta E \approx \gamma$, which gives a spread in velocities of $\delta v \approx (dv/dE)\delta E \approx \gamma/p$. As a result there is a spread in position, after time t, of about $\delta x_D \approx \delta vt \approx (\gamma/p)t = (\gamma/p)(r/v)$ and the leading edge of the packet will be rounded off by about this amount. The width of the packet, call it δx_P, is given by $\gamma\delta x_P/v \approx 1$, as follows from (10.39). The shape of the packet, as shown in Fig.10.6, will be completely altered if the spread due to dispersion is greater than the initial width of the packet, thus if $\delta x_D > \delta x_P$, or $(\gamma r/pv) > v/\gamma$, or

$$\gamma r/v > pv/\gamma = 2E_A'/\gamma. \qquad (10.41)$$

We shall next show how $\xi(r,t)$ can be evaluated in this case, using the method of stationary phase.

Consider a radial problem in which the asymptotic form of the wave function for large r is

$$\varrho \approx \frac{1}{(4\pi)^{1/2}ir}\int_0^\infty \frac{1}{(2\pi v)^{1/2}}[B(E)e^{i(pr-Et+\delta)}$$
$$-B(E)^*e^{-i(pr+Et+\delta)}]dE. \qquad (10.42)$$

For most values of E the phase of the exponential factors
is very rapidly varying; the contributions to the integral
come only from points of stationary phase. It is expedient
to change the variable of integration from E to p, since
the rapidly varying factors in the exponentials, pr±Et, are
then simple quadratic functions of the variable. The
positive exponential term is then

$$\varrho \approx [1/(4\pi)^{1/2}ir]\int_0^\infty (v/2\pi)^{1/2}B(E)e^{i(pr-Et+\delta)}dp \qquad (10.43)$$

and the condition for stationary phase is

$$(d/dp)(pr-Et+\delta)=r-v(t-\delta')=0. \qquad (10.44)$$

(The prime on δ' continues to represent differentiation
with respect to E.) Let us call the value of v given by
(10.44) v_t,

$$v_t=r/(t-\delta').$$

This is just the velocity required to reach the point r in
time t.

The phase of the negative exponential term in (10.42)
is always rapidly varying for positive t; there is no point
of stationary phase and no contribution to the integral.

The contributions to the integral in (10.43) come from
the immediate neighborhood of $p_t=Mv_t$. Expanding the phase
in a Taylor series about p_t gives

$$pr-Et+\delta=p_tr-E_tt+\delta(E_t)+(1/2)(d^2/dp^2)(pr-Et+\delta)|_{p_t}(p-p_t)^2$$

$$=\frac{Mr^2}{2(t-\delta'(E_t))}+\delta(E_t)-\frac{1}{2M}[t-\delta'(E_t)-2E_t\delta''(E_t)](p-p_t)^2,$$

where $E_t=(1/2M)v_t^2$. Thus (10.43) becomes

$$\delta \approx \frac{B(E_t)}{(4\pi)^{1/2} i r} (v_t/2\pi)^{1/2} e^{(i/2)[Mr^2/(t-\delta'(E_t))]+i\delta(E_t)}$$

$$\cdot \int_{-\infty}^{\infty} e^{-(i/2M)[t-\delta'(E_t)-2E_t\delta''(E_t)](p-p_t)^2} dp$$

$$\approx \frac{B(E_t)}{(4\pi)^{1/2} i^{3/2} r} \left[\frac{Mv_t}{t-\delta'(E_t)-2E_t\delta''(E_t)}\right]^{1/2}$$

$$\cdot e^{(i/2)[Mr^2/(t-\delta'(E_t))]+i\delta(E_t)}. \qquad (10.45)$$

The phase of the exponential in (10.45) is similar to that in (10.23). Its meaning is readily interpreted: the momentum operator applied to δ gives as the leading term $(1/i)(d/dr)\delta=[Mr/(t-\delta'(E_t))]\delta=p_t\delta$. The particle current is

$$j=v_t|\delta|^2=\frac{|B(E_t)|^2 2E_t}{4\pi r^2(t-\delta'(E_t)-2E_t\delta''(E_t))}. \qquad (10.46)$$

Since

$$\frac{dE_t}{dt}=\frac{d}{dt}\frac{Mr^2}{2(t-\delta'(E_t))^2}=-\frac{2E_t}{t-\delta'(E_t)}[1-\delta''(E_t)\frac{dE_t}{dt}],$$

$$[1-\frac{2E_t\delta''(E_t)}{t-\delta'(E_t)}]\frac{dE_t}{dt}=-\frac{2E_t}{t-\delta_t'(E)},$$

$$-\frac{2E_t}{t-\delta'(E_t)-2E_t\delta''(E_t)}dt=dE_t,$$

the total flux through the surface of a sphere of radius r is

$$4\pi r^2\int_0^\infty jdt=\int_0^\infty |B(E_t)|^2 dE_t=1,$$

as required by the conservation law.

In evaluating (10.42) we have neglected the variation of B(E) in the neighborhood of E_t. As we see from the Gaussian integral in (10.45), the contributions to the integral come from a range of momentum around p_t of order $\delta p \approx (M/t)^{1/2}$. The condition that B(E) be slowly varying is $|(dB(E_t)/dp)\delta p| \ll |B(E_t)|$, or

$$|(d\ln B(E_t)/dp)(M/t)^{1/2}| \ll 1. \qquad (10.47)$$

Applying this argument to our decay packet, (10.36), we see that the rapidly varying factor in B(E) is $1/(E_A-E-\Delta(E)-i\gamma(E))$, so

$$d\ln B(E_t)/dp=v/(E_A-E_t-\Delta(E_t)-i\gamma(E_t))$$

and the condition (10.47) can be written

$$t>>2E_t/[(E_A-E_t-\Delta(E_t))^2+\gamma(E_t)^2].$$

The critical case occurs for $E_t=E_A'$, where the condition becomes $\gamma(E_A')t>>2E_t/\gamma(E_A')$, or, since at $E_t=E_A'$ we have $v(E_A')=v_t=r/t$,

$$\gamma(E_A')r/v(E_A')>>2E_A'/\gamma(E_A'),$$

in agreement with the estimate (10.41).

For the wave function (10.36), (10.46) becomes

$$J=(1/\pi)\{\gamma(E_t)/[(E_A-E_t-\Delta(E_t))^2+\gamma(E_t)^2]\}(dE_t/dt). \qquad (10.48)$$

Thus, if the observation distance, r, is so large that the effects of dispersion have had time to have their full effect, we no longer see any sign of the exponential falloff of (10.40), but instead see a Lorentzian current distribution, (10.35), modified the factor dE_t/dt.

It is not too difficult to make a better calculation by taking into account the rapid variation of B(E) near the resonance energy. In order to simplify the algebra we shall drop terms proportional to Δ' and γ' and also take $\delta_p=0$. We rewrite (10.43) as

$$\xi\approx[1/(4\pi)^{1/2}ir]\int_0^\infty (v/2\pi)^{1/2}e^{i(pr-Et-i\ln B)}dp.$$

The asymptotic form of the integral for large r and t can be found by the saddlepoint method, a generalization of the method of stationary phase. The saddlepoint method is briefly described as follows: Write the integral as

$$\xi\approx\int_0^\infty e^{iS(p)}dp$$

and suppose there is a saddlepoint at p_s , that is, that

$$dS(p_s)/dp=0. \qquad (10.49)$$

Suppose further that the path of integration can be deformed to a contour passing through the saddle. On expanding S(p) in a Taylor series we obtain

$$\delta \approx \int_0^\infty e^{i[S(p_s)+1/2(d^2S(p_s)/dp^2)(p-p_s)^2]} dp$$

$$\approx [2\pi/(-id^2S(p_s)/dp^2)]^{1/2} e^{iS(p_s)}. \qquad (10.50)$$

In our case, with $B(E)=-V_{EA}/(E_A-E-\Delta(E)-i\gamma(E))$, (10.49) becomes

$$r-vt+iv/(E-E_A+\Delta(E)+i\gamma(E))=0. \qquad (10.51)$$

Let v_s be a root of this equation, and write $v_s=v_t+w$, where, as before, $v_t=r/t$. If we further suppose that w is small compared to v_t, (10.51) becomes

$$wt-i(v_t+w)/(E_t-E_A+\Delta(E_t)+i\gamma(E_t)+Mv_tw)=0,$$

where, again as before, $E_t=(1/2)Mv_t^2$. Multiplying through by the denominator of the last term we obtain a quadratic equation for w.

Let us call the two roots of the quadratic equation w_+ and w_-. In general, one of the roots will be small and the other large. For $v_t>v(E_A^r)$ (or $t<r/v(E_A^r)$, the arrival time of particles of resonance velocity) the root w_+ is small and the saddlepoint lies close to Mv_t in the upper half plane. For $v_t<v(E_A^r)$ (or $t>r/v(E_A^r)$) the root w_- is small and the saddlepoint lies close to Mv_t in the lower half plane. It is thus necessary to switch at (or near) $v_t=v(E_A^r)$ from a contour through the saddlepoint in the upper half plane to a contour through the saddlepoint in the lower half plane.

If the distance, r, is small compared to the critical value given by (10.41), $\gamma(E_A^r)r/v(E_A^r)<<2E_A^r/\gamma(E_A^r)$, it is found that the contour in the lower half plane passes below the pole at $E_A^r-i\gamma(E_A^r)$. In this case, in deforming the contour one picks up the residue at the pole, and instead of (10.50) one obtains the sum of (10.50) and (10.39). Thus for $t<r/v(E_A^r)$ one receives a wave of velocity $v_t=r/t$, while for $t>r/v(E_A^r)$ one receives the sum of two waves, one of velocity r/t and the other of the resonance velocity. This change in the form of the result is the analogue of the discontinuity shown in the approximation (10.39).

In the opposite limit, $\gamma(E_A')r/v(E_A') >> 2E_A'/\gamma(E_A')$, the deformed contour in the lower half plane passes above the pole, and both for small and large t there is only a single wave of velocity r/t, in agreement with (10.45).

Since only some straightforward algebra is involved in the evaluation of (10.50), we shall skip the details and show some results in the form of plots of the particle current as a function of time which would be seen at three different distances. We have taken $E_A'/r=500$ (remember that the decay rate $\Gamma=2\gamma(E_A')$. Fig.10.7 is for a distance such that particles of the resonant velocity take 0.2 mean lives to arrive, $\Gamma r/v(E_A')=0.2$, Fig.10.8 is for $\Gamma r/v(E_A')=20$ and Fig.10.9 for $\Gamma r/v(E_A')=20000$, larger than the critical value, 4000, given by (10.41).

For the first two cases we also show, in the lighter curve, the simple exponential decay of (10.40). The oscillations about the exponential curve are, of course, due to interference between the v_t and $v(E_A')$ waves. The amplitude of the oscillations shows the amplitude of the v_t wave, while the oscillation period gives the time, according to the uncertainty relation, required to resolve the energy difference between the waves. The saddlepoint evaluation of the integral is not too accurate in the immediate neighborhood of $v_t=v(E_A')$: the values from the upper and lower contours at $v_t(E_A')$ do not quite agree, as is shown by the discontinuity of the curves at this point. As r approaches the critical value, (10.41), the disagreement becomes larger, although it is not difficult to interpolate between the two curves by eye.

In the third graph, that for very large r, the curves from the upper and lower saddlepoints fit smoothly together at the maximum of the curve. The lighter curve is that given by (10.48). Here, for $\Gamma r/v(E_A')$ five times the critical value, the saddlepoint curve is seen to be

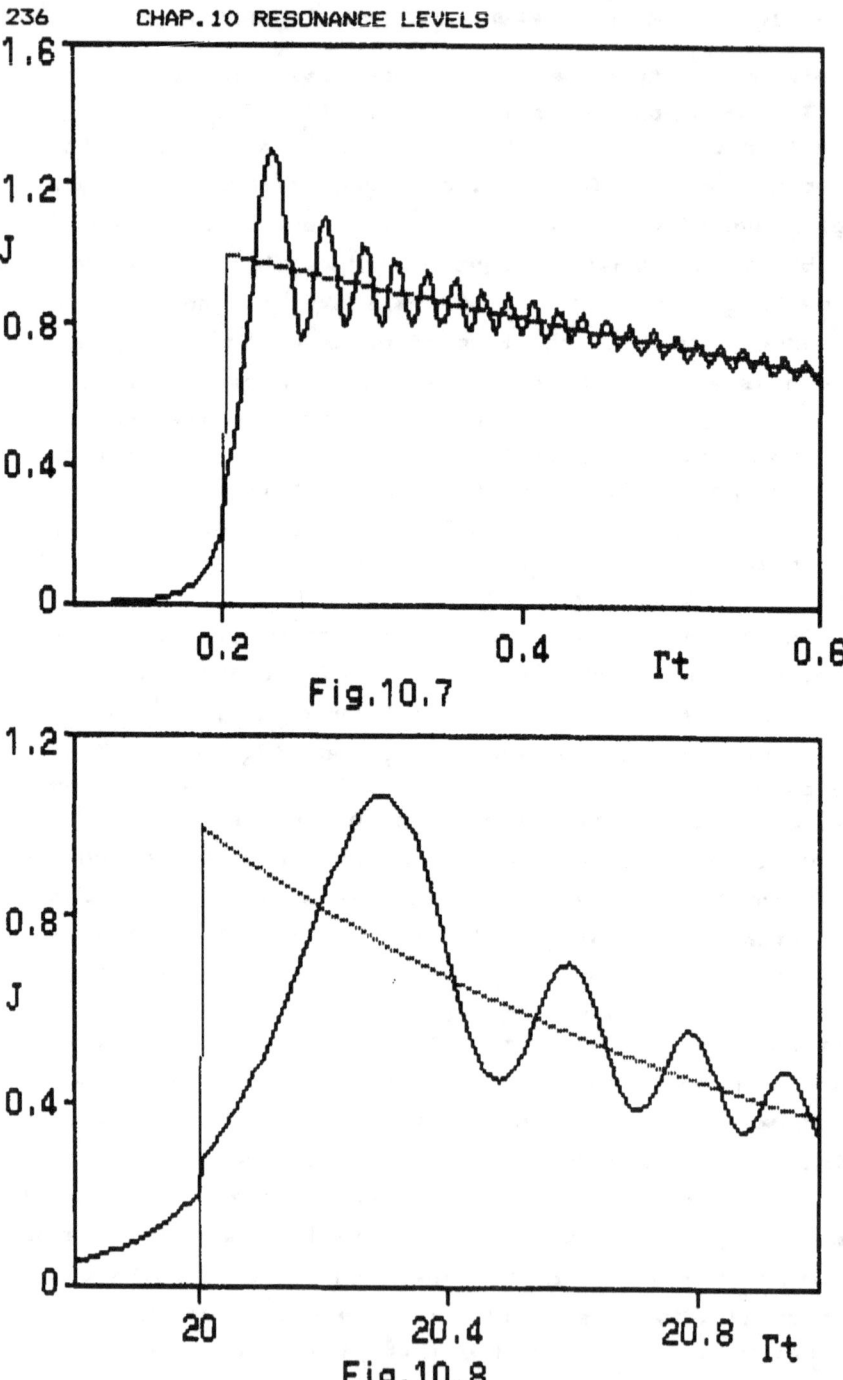

Fig.10.7

Fig.10.8

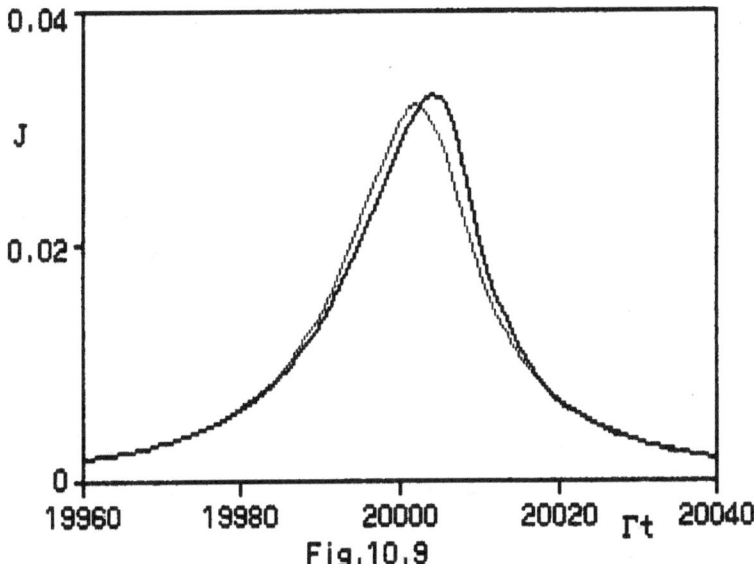

Fig.10.9

approaching the Lorentzian form.   There is a point here to be remarked on. In (10.43) only the product B(E)exp(iδ) is defined, not the two factors separately.   In calculating the light curve we have taken B(E) real and δ=δ$_R$, as given by (10.34). Thus we have treated δ$_R$ on the same footing as δ$_P$. The meaning of the δ' correction to the arrival time has already been explained for δ$_P'$. The reason for the δ$_R'$ correction is not immediately obvious. However, we see from the first two graphs that the rounding of the signal caused by dispersion results in a delay of the maximum which increases with r. The δ$_R'$ correction gives the limiting value of this delay. With δ$_R$ given by (10.34) we have δ$_R'$(E$_A'$)=1/γ(E$_A'$)=2/Γ, or a delay of two mean lives.

   Another interesting case is that for fixed r as t→∞. When E$_t$<<E$_A'$ the wave function is the sum of the decay amplitude, (10.39), and the amplitude (10.45) with B(E)=-V$_{EA}$/(E$_A$-Δ(0)). Since the scattering length approximation holds as E$_t$→0, the δ' and δ''terms should be dropped and r replaced by r-a. After the decay amplitude dies out we are left with

$$\psi \approx \emptyset_B \emptyset_b \frac{1}{(4\pi)^{1/2} i^{3/2} r} \frac{V_{E_t A}}{(E - \Delta(0))} (Mv_t/t)^{1/2} e^{iM(r-a)^2/2t}$$

and a density

$$|\psi|^2 = (1/4\pi r^2) |V_{E_t A}/(E_A - \Delta(0))|^2 (Mv_t/t). \qquad (10.52)$$

To show the velocity dependence explicitly, use (10.17) to write $V_{E_t A}$, as $E_t \to 0$, in terms of V_{0A} the momentum normalized matrix element for $p=0$. This gives

$$|\psi|^2 = |V_{0A}/(E_A - \Delta(0))|^2 M^3 v_t^2 / r^2 t$$

$$= |V_{0A}/(E_A - \Delta(0))|^2 (M/t)^3 (1 - a/r)^2.$$

This result is easily related to the discussion of the preceding section. Aside from the scattering length correction, this is the same as (10.26) with $c_0 = V_{0A}/(E_A - \Delta(0))$.

The current through the surface of a sphere of radius r is

$$J = 4\pi r^2 v_t |\psi|^2 = 4\pi |c_0|^2 [M(r-a)]^3/t^4.$$

The r^3/t^4 dependence looks odd at first sight. A check is provided by the obvious conservation law: the integral of the current from t to ∞ is

$$\int_t^\infty J dt = (4\pi/3) |c_0|^2 [M(r-a)]^3/t^3 = \int_a^r |\psi|^2 dr,$$

just the probability a particle is within the sphere at time t.

The use of the stationary phase method to evaluate (10.36) seems dubious when $M(r-a)^2/2t$ (or $E_t t$) is less than 1. However, in this case (10.36) can be evaluated by deforming the contour to the negative imaginary axis, as was done in (10.19) for (10.11), and one verifies that (10.52) is still correct. It may be noted that both terms in (10.36) contribute to the answer, not just the first.

Even if there is no dispersion, as in the emission of a photon, there will be a residual signal of the type we

have been discussing, after the exponential decay has died
out, due to the presence in the wavepacket of wavelenghths
longer than ct.

SEC.10.IV. THE SCATTERING CROSS SECTION

The phase shift of our stationary state wavefunction,
(10.32), is $\delta(E)=\delta_R(E)+\delta_P(E)$, with $\delta_R(E)$ given by (10.34).
The resonance phase shift increases from ≈ 0 below the
resonance to $\pi/2$ at E'_A to $\approx\pi$ above. The rationale for
this behavior is easily seen from a simple potential model.
Consider s wave scattering by an attractive square well
potential. If the well is infinitely deep we obtain a bound
state each time the phase of the wavefunction at the radius
of the well changes by π. If the well has a depth V and V
is increased, the phase shift of the zero-energy solution
increases by π each time a new state becomes bound. If we
add a repulsive barrier at the edge of the well we may
produce a resonance level at a positive energy; again the
phase shift increases by π as the energy increases through
the resonance, a special example of the more general
treatment given in this chapter.

We started the chapter with a representation having a
continuous spectrum ϵ and a state A, and ended with just
the continuous spectrum E. The presence of the missing
state is still indicated, however, by the increase of the
phase shift by π in passing through the resonance.

The scattering amplitude for the scattering of b by B
is given by

$$f_0=(1/2ip)(e^{2i\delta(E)}-1). \qquad (10.53)$$

If δ_P varies little over the width of the resonance
the amplitude has a value determined by δ_P below the
resonance and returns to approximately the same value
above. If $\delta_P=0$ the scattering cross section has
approximately the Lorentzian shape (Fig.10.10a). If $\delta_P<0$
and $|\delta_P|<90°$ the phase shift starts out negative and, as the
energy increases, passes through 0 before reaching the

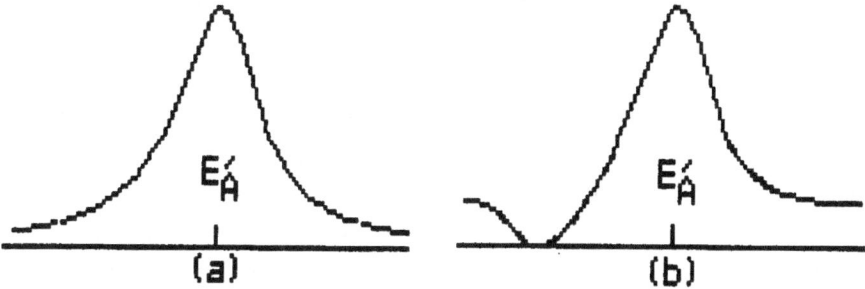

Fig.10.10

resonance energy. At this point the scattering amplitude
is 0, so the cross section will first dip to zero, then
rise to a maximum beyond the resonance energy, before
dropping again (Fig.10.10b). For $\delta_P > 0$ the dip occurs above
the resonance energy. For the anomalous case, $\delta_P = \pm 90°$, we
would get an inverted Lorentzian curve, the cross section
dipping to 0 at the resonance energy.

Using $\delta(E) = \delta_P(E) + \delta_R(E)$, we can write (10.53) in the
form

$$f_0 = (1/2ip)[(e^{2i\delta_P(E)} - 1) + e^{2i\delta_P(E)}(e^{2i\delta_R(E)} - 1)],$$

or, using (10.33), as

$$f_0 = (1/2ip)(e^{2i\delta_P(E)} - 1) + (1/p)e^{2i\delta_P(E)} \gamma(E)/(E_A - E - \Delta(E) - i\gamma(E)).$$

The first term is commonly called potential scattering.

If $\delta_P(E) = 0$ the scattering cross section is

$$\sigma = 4\pi|f_0^2| = (4\pi/p^2)\gamma(E)^2/[(E_A - E - \Delta(E))^2 + \gamma(E)^2]. \qquad (10.54)$$

If $\Delta'(E)$ and $\gamma'(E)$ can be neglected the cross section,
written in terms of the decay constant, Γ, takes the
Lorentzian form,

$$\sigma = (\pi/p^2)\Gamma^2/[(E_A' - E)^2 + (\Gamma/2)^2].$$

SEC.10V SCATTERING WITH SEVERAL DECAY CHANNELS

The scattering problem can also be solved for the case
in which there are several decay channels for the state A,
$$A \rightarrow B_j + b_j, \quad j=1 \text{ to } n.$$
The stationary state spectrum will now have an additional
n-fold degeneracy. If we re-solved the decay problem, the
Fourier transform of the wave function would give us a
particular linear combination of the n solutions, not the
solutions separately. What we would like is to get n
solutions, each categorized by there being just particles
of one channel in the incoming wave. We accomplish this by
solving the stationary state problem directly.

The equations (10.01) are modified as follows: The
second equation is replaced by n equations, one for each
channel. The energy thresholds of the channels, $E_{j,th}$ will
in general be different. If ϵ_j is to run from 0 to ∞, ϵ in
the second equation is to be replaced by $E_{j,th} + \epsilon_j$. And,
since for the stationary state solution the time dependence
of all the c's is $\exp(-iEt)$, idc/dt is replaced by Ec. If
we let $E_j = E - E_{j,th}$ we then have in place of (10.01)

$$(E-E_A)c_A = \Sigma_j \int_0^\infty V_{A\epsilon_j} c_{\epsilon_j} d\epsilon_j,$$
$$(E_j - \epsilon_j)c_{\epsilon_j} = V_{\epsilon_j A} c_A.$$

To have the incoming particle in the i'th channel, we
take
$$c_{\epsilon_j} = \delta_{j,i} \delta(E_i - \epsilon_i) + b_{\epsilon_j}.$$
Then
$$b_j = -V_{\epsilon_j A} c_A / (\epsilon_j - E_j)$$
and
$$c_A = -V_{A\epsilon_i} / (E_A - E - \Sigma(E)), \tag{10.55}$$
where now
$$\Sigma(E) = \Sigma_j \int_0^\infty |V_{\epsilon_j A}|^2 / (\epsilon_j - E_j) d\epsilon_j, \tag{10.56}$$

and we specify that E approaches the real axis from

above, both in (10.55) and (10.56). As we shall soon see, this choice results in the desired boundary conditions: only outgoing waves in the channels other than i. We then have

$$\Sigma(E) = \Delta(E) + i\gamma(E),$$

with $\gamma(E)$ now given by

$$\gamma(E) = \Sigma_j \gamma_j$$

and

$$\gamma_j = \pi |V_{\epsilon_j A}|^2$$

provided emission of particle b_j is energetically possible (i.e. $E_j > 0$), else $\gamma_j = 0$.

In place of (10.29) we now have

$$\xi(.r_j..) = \Sigma_j \int c_{\epsilon_j} \emptyset_{\epsilon_j}(r_j) d\epsilon_j + c_A \emptyset_A$$

$$= \emptyset_{\epsilon_i} + \Sigma_j \int \frac{\emptyset_{\epsilon_j} V_{\epsilon_j A} V_{A \epsilon_i}}{(\epsilon_j - E_j - i0)(E_A - E - \Delta(E) - i\gamma(E))} d\epsilon_j - \frac{\emptyset_A V_{A \epsilon_i}}{(E_A - E - \Delta(E) - i\gamma(E))}.$$

If we insert the asymptotic forms, (10.28), of the radial functions for large r_j, we can evaluate the integrals in terms of the residues at the poles, as we did for (10.31). The result can be written, aside from a constant factor, in the form

$$\xi(.r_j..) \approx \frac{\emptyset_{B_i} \emptyset_{b_i} \sin(p_i r_i)}{v_i^{1/2} p_i r_i} + \Sigma' \frac{\emptyset_{B_j} \emptyset_{b_j} f_{ji} e^{p_j r_j}}{v_j^{1/2} r_j}, \qquad (10.57)$$

with

$$f_{ji} = \delta_{j,i}(1/2ip_i)(e^{2i\delta_{Pi}} - 1)$$
$$+ (1/p_i) e^{i(\delta_{Pi} + \delta_{Pj})} \pi V_{\epsilon_j A} V_{A \epsilon_i} / (E_A - E - \Delta(E) - i\gamma(E)).$$

The prime on the summation sign in (10.57) means that the sum is to be taken only over those j for which $E_j > 0$.

The constant by which ξ was multiplied was chosen so the first term on the right in (10.57) is the s wave part of

$$(1/v_i)^{1/2} e^{i\underline{p}_i \cdot \underline{r}_i},$$

an incoming plane wave normalized to unit current. The

remaining terms in (10.57) give the outgoing scattered waves. With this normalization, the total scattering cross section, σ_{ji}, for scattering into the j channel is just the current through the surface of a sphere of radius r_j and is thus

$$\sigma_{ji}=4\pi|f_{ji}|^2.$$

For j not equal to i,

$$\sigma_{ji}=(4\pi/p_i^2)\gamma_i(\epsilon_i)\gamma_j(\epsilon_j)/[(E_A-E-\Delta(E))^2+\gamma(E)^2]. \qquad (10.58)$$

If $\delta_{Pi}=0$, σ_{ii} is also given by (10.58), which differs from (10.54) in that $\Delta(E)$ and $\gamma(E)$ are now the sums of the contributions from all the channels. If $\Delta'(E)$ and $\gamma_j'(\epsilon_j)$ can be neglected, (10.58) can be written

$$\sigma_{ji}=(\pi/p_i^2)\Gamma_j\Gamma_i/[(E_A'-E)^2+\Gamma^2/4], \qquad (10.59)$$

with $\Gamma_j=2\gamma_j(E_A'-E_{j,th})$ and $\Gamma=2\gamma(E_A')$.

The resonance scattering formualae (10.58) and (10.59) have been derived for s wave scattering and no spin-orbit interactions. It is not too difficult to remove these restrictions. If the scattering is from a wave of angular momentum l, the cross sections are multiplied by a factor 2l+1, e.g. (10.59) becomes

$$\sigma_{ji}=[\pi(2l+1)/p_i^2]\Gamma_j\Gamma_i/[(E_A'-E)^2+\Gamma^2/4]. \qquad (10.60)$$

If there are spin interactions and the spins in the incident channel are j_{b_i} and j_{B_i} and of the resonance level J, an argument similar to that of Sec.9I shows that,for unpolarized beams, (10.60) should be multiplied by the fraction of levels in a state J, $(2J+1)/[(2j_{b_i}+1)(2j_{B_i}+1)(2l+1)]$, so

$$\sigma_{ji}=\frac{\pi(2J+1)}{(2j_{b_i}+1)(2j_{B_i}+1)p_i^2}\frac{\Gamma_j\Gamma_j}{(E_A'-E)^2+\Gamma^2/4}. \qquad (10.61)$$

The way the weight factors appear in (10.61) can also be deduced by a detailed balancing argument. The scattering process can be envisaged as taking place in two steps,

first the absorption of b_i to produce A, with an absorption
cross section $\sigma_i = \Sigma_j \sigma_{ji}$, then the decay of A, with a
probability Γ_j / Γ of decaying into channel j. Consider
equilibrium in an energy interval δE, with $\delta E \gg \Gamma$ but still
small enough that p_i doesn't vary appreciably over δE.
Equating the rate of decay of A into channel i to the rate
of formation of A from channel i gives

$$(2J+1)\Gamma_i = (2j_{b_i}+1)(2j_{B_i}+1)v_i \langle \sigma_i \rangle 4\pi p_i^2 (dp_i/dE_i)\delta E/(2\pi)^3,$$

or

$$\langle \sigma_i \rangle = 2\pi^2 [(2J+1)/(2j_{b_i}+1)(2j_{B_i}+1)]\Gamma_i/(p_i^2 \delta E). \quad (10.62)$$

Here $\langle \sigma_i \rangle$ is the average cross section,

$$\langle \sigma_i \rangle = \int_{-\infty}^{\infty} \sigma_i \, dE/\delta E. \quad (10.63)$$

Inserting the σ_i given by (10.61) in (10.63) and
doing the integral, we obtain a $\langle \sigma_i \rangle$ which is just the same
as that, (10.62), obtained from the detailed balancing
argument.

PHYSICAL REVIEW C   VOLUME 14, NUMBER 2   AUGUST 1976

A simple nuclear model*

R. Serber

Columbia University, New York, New York 10027

(Received 26 January 1976)

A nuclear model is considered in which the nuclear forces are taken to be a Yukawa potential acting between nucleons, such as would be produced by a neutral scalar meson, and a repulsive core. The model is very similar to the Fermi-Thomas model of the atom. It is first worked out in the absence of the Coulomb interaction, and it is shown how the binding energy, surface energy, and shape of the nuclear density at the surface of the nucleus are determined. The Coulomb interaction is then included and the model is worked out for a heavy nucleus, the parameters of the model being chosen to give the correct radius and binding energy. One predicts the right value of Z/A, and obtains a charge density distribution not too different from that deduced from electron scattering experiments.

NUCLEAR STRUCTURE Model, neutral scalar meson interaction plus repulsive core. Calculated binding energy, charge, neutron, and proton spatial distributions.

INTRODUCTION

In the course of following the work of Lee and Wick[1] on abnormal nuclear matter, it occurred to the author to apply the same methods to normal nuclear matter and to develop a model describing the ground-state properties of ordinary nuclei. The model should incorporate the principle features of the forces which bind the nucleons, but be sufficiently simple so that all calculations can be carried through explicitly with little labor. The model is very similar to the Fermi-Thomas model of the atom, but its solution is even simpler, at least in the approximation in which the energy is divided into volume and surface terms.

At first sight the nucleon-nucleon forces derived from scattering experiments in the several-hundred MeV range do not seem at all simple, but a good deal of the complexity disappears when the forces are averaged over the nuclear ground state. For example, the long-range forces due to single pion exchange are a tensor force and a spin-dependent central force; the former vanishes for a spherical nucleus, the latter for a nucleus with spin zero. The result of the averaging leaves primarily a short-range spherical force (presumably due to scalar mesons) and a repulsive core (presumably due to vector mesons).

I. NUCLEAR MODEL WITHOUT COULOMB FIELD

We shall take the potential between a pair of nucleons to be a simple Yukawa interaction, so that the potential energy is

$$U = -\frac{1}{2}\frac{g^2}{4\pi}\int\rho(r)\frac{e^{-\mu|\vec{r}-\vec{r}'|}}{|\vec{r}-\vec{r}'|}\rho(r')\,d\vec{r}'\,d\vec{r}, \quad (1)$$

where $\rho(r)$ is the density of nucleons. If we write

$$\phi(r) = \frac{g}{4\pi}\int\frac{e^{-\mu|\vec{r}-\vec{r}'|}}{|\vec{r}-\vec{r}'|}\rho(r')\,d\vec{r}', \quad (2)$$

the quantity ϕ satisfies the Yukawa equation,

$$\Delta\phi - \mu^2\phi = -g\rho. \quad (3)$$

As in electrostatics ($\mu = 0$), the advantage of (3) is twofold. It may be easier to solve the differential equation than do the integral. And ϕ and ρ may be codetermined, as in electrostatic problems with conductors. The interaction energy is

$$V_{\text{int}} = -g\int\rho\phi\,d\vec{r}. \quad (4)$$

Note that (4) differs from (1) by a factor 2; in fact, (1) can be written

$$U = \int[\tfrac{1}{2}(\text{grad}\phi)^2 + \mu^2\phi^2]\,d\vec{r} + V_{\text{int}}. \quad (5)$$

The first term, the field energy, is analogous to the field energy $\frac{1}{2}E^2$, $E = -\text{grad}\Phi$, of electrostatics (plus an additional μ^2 term), and, by partial integration of the $(\text{grad}\phi)^2$ term and use of (3), it can easily be seen that the field energy just equals $-\frac{1}{2}V_{\text{int}}$.

Equation (5) gives the field and interaction energies for a neutral scalar meson field (σ meson). The idea of using ϕ as a variable was the one borrowed from Lee and Wick, though the author should have learned it from Yukawa.

For the kinetic energy of the nucleons, including

the effect of the repulsive core, we shall use the formula for the energy of a degenerate hard-sphere Fermi gas in the form given by Bohr and Mottelson,[2]

$$T = \int \frac{3}{5} \frac{1}{2M} \frac{p_F^2}{(1 - a p_F)^2} \rho \, d\vec{r}, \tag{6}$$

where $a = (5/3\pi) r_c$, r_c is the core radius, and, for equal numbers of protons and neutrons, the Fermi momentum is related to the density by

$$\rho = \frac{2}{3\pi^2} p_F^3. \tag{7}$$

According to Bohr and Mottelson, (6) gives a good representation of the kinetic energy over the range of variables for which we will be using it, $a p_F < \frac{1}{2}$. The form of the correction for finite core radius is entirely reasonable, leading, as it does, to a limiting density determined by $p_F = 1/a$.

The total energy is thus

$$E = \int \left[\frac{1}{2} (\mathrm{grad}\phi)^2 + \frac{3}{5} \frac{1}{2M} \frac{p_F^2}{(1 - a p_F)^2} \rho \right.$$
$$\left. + \frac{1}{2} \mu^2 \phi^2 - g\phi\rho \right] d\vec{r}. \tag{8}$$

The energy E is given in terms of two variables, p_F and ϕ. The mathematical problem is to minimize E with respect to variations of p_F and ϕ, subject to the condition that the total number of particles,

$$A = \int \rho \, d\vec{r}, \tag{9}$$

remains constant.

As a simple example consider the case of a nucleus of constant density with no repulsive core, $a = 0$. For constant ρ the solution of (3) is

$$\phi = \frac{g}{\mu^2} \rho, \tag{10}$$

and the binding energy per particle is

$$\frac{E}{A} = \frac{3}{5} \frac{p_F^2}{2M} - \frac{1}{2} g\phi = \frac{3}{5} \frac{p_F^2}{2M} - \frac{1}{2} \frac{2}{3\pi^2} \frac{g^2}{\mu^2} p_F^3. \tag{11}$$

If we set the derivative of E/A with respect to p_F equal to zero we obtain a solution

$$p_F = \frac{4}{5} \frac{3\pi^2}{2} \frac{1}{2M} \frac{\mu^2}{g^2}. \tag{12}$$

This, however, is a maximum, rather than a minimum of E/A. As shown in Fig. 1(a), for small p_F (low density) the kinetic energy dominates, for large p_F the potential energy. The physical situation is illustrated by Fig. 1(b). Imagine the Fermi gas in a cylinder being compressed by a piston.

At low density the gas exerts a positive pressure on the piston. At the critical density, given by (12) and (7), we reach a condition of unstable equilibrium, the attractive forces between nucleons just balancing the Fermi pressure, and the pressure on the piston is zero. If the piston is pushed a little further the system collapses to a droplet on the cylinder floor. In our approximation the density would be infinite; the author first learned from Lee and Wick that this is not so if relativistic formulas are used for the kinetic and potential energies. There is then a solution of the problem, but with density and binding energy larger than that of ordinary nuclei. This observation is the basis of the theory of abnormal nuclear matter. With a repulsive core, the curve turns up, as indicated by the dotted line, and the energy has a minimum.

The variation problem stated in connection with (8) can be treated by introducing a Lagrange multiplier to take care of the condition (9); we write

$$E + \lambda A = \int \left[\frac{1}{2} (\mathrm{grad}\phi)^2 + \frac{3}{5} \frac{1}{2M} \frac{p_F^2}{(1 - a p_F)^2} \rho \right.$$
$$\left. + \frac{1}{2} \mu^2 \phi^2 - g\rho\phi + \lambda\rho \right] d\vec{r}, \tag{13}$$

and require

$$\delta(E + \lambda A) = 0. \tag{14}$$

The significance of the Lagrange multiplier λ is made evident by (14),

$$\frac{\partial E}{\partial A} = -\lambda. \tag{15}$$

Before proceeding with the variational equations, it is convenient to simplify the form of (13) by introducing dimensionless measures of the quantities involved. We shall measure r in terms of the meson Compton wavelength (range of the Yukawa force), $r = r_s/\mu$. For p_F we use the unit which appears in (12)

$$p_F = \frac{3\pi^2}{2} \frac{1}{2M} \frac{\mu^2}{g^2} p_s, \tag{16}$$

so that in this unit the maximum of Fig. 1(a) occurs at $p_s = \frac{4}{5}$. To find the appropriate unit for the field variable, we note that $g\phi$ is an energy, as is $p_F^2/2M$; thus[3]

$$g\phi = \left(\frac{3\pi^2}{2} \right)^2 \left(\frac{1}{2M} \right)^3 \frac{\mu^4}{g^4} \phi_s. \tag{17}$$

Also we take $a p_F = a_s p_s$, $\rho_s = p_s^3$. In the ensuing, the subscript s will be dropped, and it is to be understood that all quantities are in their scaled measure unless it is explicitly stated otherwise. The change to scaled quantities is accomplished simply by setting g, μ, $2M$, and the factor $(2/3\pi^2)$ of Eq. (7) equal to unity. Thus the scaled Eq. (13) reads

$$E + \lambda A = \int \left[\tfrac{1}{2}(\operatorname{grad}\phi)^2 + \frac{3}{5}\,\frac{p^5}{(1-ap)^2} + \tfrac{1}{2}\phi^2 \right.$$
$$\left. - \phi p^3 - \lambda p^3 \right] d\vec{r}. \tag{18}$$

II. SURFACE LAYER

The basic assumption we shall make to simplify the solution of (14) is that the range of forces, $1/\mu$, is very small compared to the nuclear radius R. The nucleons will then have an interior of constant density, surrounded by a thin surface layer of thickness T, of order $1/\mu$.

It will be observed that the scaling equations, (16) and (17), depend only on g^2/μ^2. In consequence, the properties of the nuclear interior (density, volume binding energy) depend only on this ratio, and on a. This is a consequence of (10), which gives, for the unscaled quantities, $g\phi = (g^2/\mu^2)\rho$. The surface energy, however, is expressly proportional to $1/\mu$. To see this, think of the simplest description of the origin of a surface energy. The nucleons in the interior are surrounded on all sides by neighbors. The nucleons in a thin layer, T, at the surface have neighbors only on one side, and thus only half the binding energy. If the binding energy of an interior nucleon is $-\lambda$, the surface energy is $\tfrac{1}{2}\lambda$ times the number of nucleons in the surface layer, or

$$E_{\text{surf}} = \tfrac{1}{2}\lambda\rho TS, \tag{19}$$

where S is the surface area. The unscaled quantities λ and ρ depend on g^2/μ^2, but T is proportional to $1/\mu$.

The condition $T \ll R$ allows us to neglect the curvature of the surface, and reduces the variational problem to a one-dimensional case. If x

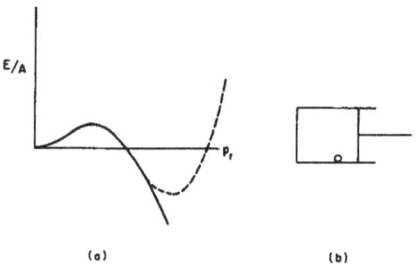

(a)　　　　　(b)

FIG. 1. E/A for the case $\rho=$ const. The dotted curve results from adding a repulsive core.

is normal to the surface, (18) becomes

$$E + \lambda A = S \int \mathcal{L}\,dx, \tag{20}$$

$$\mathcal{L} = \frac{1}{2}\left(\frac{d\phi}{dx}\right)^2 - V, \tag{21}$$

$$-V = \frac{3}{5}\,\frac{p^5}{(1-ap)^2} + \tfrac{1}{2}\phi^2 - \phi p^3 + \lambda p^3. \tag{22}$$

Note that, according to (20), $E = -\lambda A + \gamma S$, with γ some constant. There has been a subtle change in the meaning of λ, which is now the volume binding energy. This is compatible with (15) because in the one-dimensional case the surface area is constant, $\partial S/\partial A = 0$.

The variation problem is now

$$\delta \int \mathcal{L}\left(\frac{d\phi}{dx}, \phi, p\right) dx = 0, \tag{23}$$

that is, we are required to minimize the surface energy. The boundary conditions to be satisfied by ϕ are $\phi(\infty) = 0$, $\phi(-\infty) = $ const, i.e., $d\phi/dx|_{-\infty} = 0$.

We have used the notation $\mathcal{L}$ for the integrand in (20) to bring out the analogy with the motion of a particle of unit mass in a potential V. The field ϕ is analogous to the coordinate of the particle, x to the time.

The variational equations (Lagrangian equations of motion) are

$$\frac{d^2\phi}{dx^2} = -\frac{\partial V}{\partial \phi} = \phi - p^3, \tag{24}$$

that is, the one-dimensional form of (3), and

$$\frac{\partial V}{\partial p} = 0, \tag{25}$$

which, on carrying out the differentiation, gives

$$\frac{p^2}{(1-ap)^2}\left[1 + \frac{2}{5}\,\frac{ap}{(1-ap)}\right] - \phi = -\lambda. \tag{26}$$

The meaning of (26) is immediately obvious if we consider the limit $a = 0$, when it becomes $p^2 - \phi = -\lambda$. It is illustrated in Fig. 2. The potential in the nucleus is $-\phi$, and the Fermi sea is filled to a level λ below zero energy, just as in the case of the Fermi-Thomas atom. The form of the kinetic energy term given by (26) includes the $1/(1-ap)^2$ factor due to the repulsive core, and another factor, arising from differentiating the denominator of the expression for the kinetic energy density, which can be described as producing an effective nucleon mass, $M^*/M = 1/\{1 + (2/5)[ap/(1-ap)]\}$. For the range of variables we use, M^*/M varies between 0.74 and 1.

In terms of the variable

$$y = 1/(1 - ap) \tag{27}$$

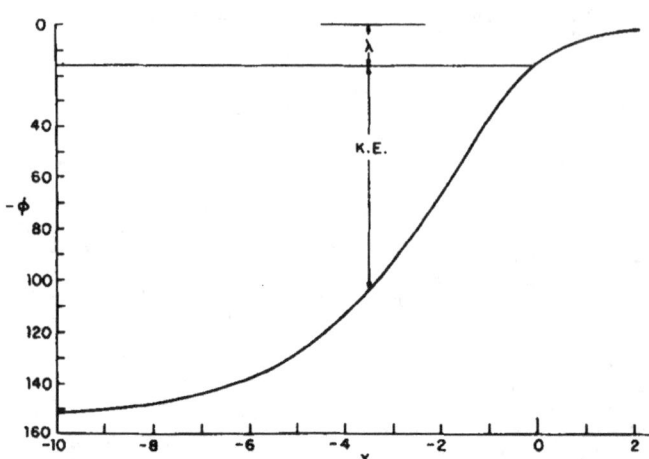

FIG. 2. Fermi-Thomas-like diagram for the example of Sec. IV. The Fermi sea is filled to the level indicated by the horizontal line.

(26) can be rewritten

$$\tfrac{1}{6}p^3y^2(1+\tfrac{1}{2}y)-\phi=-\lambda . \qquad (28)$$

Equation (28) gives ϕ explicitly as a function of p. We can imagine inverting this relation and expressing p and V as functions of ϕ. Equation (24) can be immediately integrated in terms of the "energy integral" (to follow the dynamical analogy), which is a constant,[4]

$$\epsilon = \frac{1}{2}\left(\frac{d\phi}{dx}\right)^2+V , \qquad (29)$$

$$\frac{dx}{d\phi}=\pm\frac{1}{[2(\epsilon-V)]^{1/2}} , \qquad (30)$$

$$x=\pm\int\frac{d\phi}{[2(\epsilon-V)]^{1/2}} . \qquad (31)$$

We must now discuss the form of V to see how the boundary conditions can be satisfied. We see from Fig. 2 that for $\phi<\lambda$, $p=0$; in this region $V=-\tfrac{1}{2}\phi^2$, as shown in Fig. 3. It is evident from (22) that $V\to-\infty$ as ϕ grows large. But we also know, at least for not too large a and λ, that V must have a minimum, reflecting the condition of unstable equilibrium illustrated in Fig. 1.[5] V therefore has a maximum at some value ϕ_i.

First let us consider the boundary condition $\phi(\infty)=0$. In the dynamical analogy, this means the particle approaches the origin as $t\to\infty$. This condition requires, and will be satisfied, if the energy $\epsilon=0$. The other boundary condition, $d\phi/dx|_{-\infty}=0$, requires that the velocity go to zero at

large negative t. If $V(\phi_i)<0$ the magnitude of the velocity reaches a minimum at ϕ_i, but then increases indefinitely, leading to no solution. If $V(\phi_i)>0$ there is a solution, physically meaningful for the one-dimensional case, in which ϕ starts at zero, increases to a value a little less than ϕ_i, and returns again to zero. This represents a slab of finite thickness, with a surface layer at either end. Our case, that of a semi-infinite slab, is achieved in the case $V(\phi_i)=0$, the particle starting at ϕ_i at infinite negative time and reaching zero at infinite positive time.

The unique specification $V(\phi_i)=0$ is not gratuitous. It must be remembered that V depends on the parameter λ. The condition is thus in the nature of an eigenvalue condition; satisfying it determines the value of the volume binding energy λ.

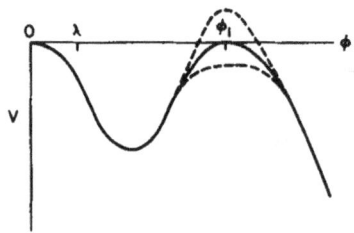

FIG. 3. V as a function of ϕ.

For $\epsilon = 0$, the solution of the "equation of motion," (31), becomes

$$x = -\int_\lambda^\phi \frac{d\phi}{(-2V)^{1/2}} ,\qquad (32)$$

where we have taken $x = 0$ at the point $\phi = \lambda$, that is, the point where p reaches zero. Since $\epsilon = 0$ means $\frac{1}{2}(d\phi/dx)^2 = -V$ according to (29), the energy equation (20) becomes

$$E + \lambda A = S \int (-2V) dx = S \int_0^{\phi_i} (-2V)^{1/2} d\phi ,\qquad (33)$$

on using (30).

Since (26) gives ϕ explicitly in terms of p, rather than vice versa, it is more convenient to write (32) as

$$x = -\int_0^p \frac{1}{(-2V)^{1/2}} \frac{d\phi}{dp} dp , \quad x < 0 ,\qquad (34)$$

which gives x as a function of p, and thus the density $\rho = p^3$, as a function of x. For $x > 0$, $p = 0$, and

$$\phi = \lambda e^{-x} ,\qquad (35)$$

as follows immediately from (24).

The expression for the energy can be written

$$E + \lambda A = S \left[\int_0^{p_i} (-2V)^{1/2} \frac{d\phi}{dp} dp + \frac{1}{2}\lambda^2 \right] ,\qquad (36)$$

p_i being the value of p for which $\phi = \phi_i$. The term $\frac{1}{2}\lambda^2$ comes from the integration from $\phi = 0$ to $\phi = \lambda$.

III. PROPERTIES OF THE NUCLEAR INTERIOR

The properties of the nuclear interior, the volume energy λ, and the density $\rho_i = p_i^3$, are determined by (28) and the condition $V(\phi_i) = 0$. For $p = p_i$, Eq. (28) takes the form

$$\tfrac{4}{3} p_i^2 y_i^2 (1 + \tfrac{2}{3} y_i) - \phi_i = -\lambda ,\qquad (37)$$

while $V(\phi_i) = 0$ can be written

$$\tfrac{2}{3} p_i^2 y_i^2 - \tfrac{1}{2}\phi_i = -\lambda .\qquad (38)$$

In writing (38) we have used the fact that $\partial V/\partial \phi = 0$ for $\phi = \phi_i$, hence $\phi_i = p_i^3$, according to (24).

Equation (38) is the statement that the mean energy of a particle in the nuclear interior is just equal to $-\lambda$ (averaging the interior energy vertically in Fig. 2 gives the same answer as reading horizontally), evidently a necessary condition for the self-consistency of the model. This condition could have been written down directly, without going through the argument of Sec. II that $V(\phi_i) = 0$.

Putting $\phi_i = p_i^3$ and subtracting (38) from (37) we find

$$p_i = \tfrac{4}{3} y_i^3 = \frac{4}{5} \frac{1}{(1 - ap_i)^3} .\qquad (39)$$

This relationship is illustrated in Fig. 4, where p and $\frac{4}{5}/(1 - ap)^3$ are plotted against p. For $a = 0$ there is one root at $p = \frac{4}{5}$, the root of Fig. 1. As a increases the root moves to the right, and a second root appears at larger p. The larger root is p_i. For a larger than $135/1024 = 0.132$ there is no solution.

Using (39) and $\phi_i = p_i^3$, (38) can be written

$$\tfrac{2}{3} y_i^3 - y_i^2 = \lambda / \tfrac{3}{5} p_i^2 .\qquad (40)$$

Equation (40) has been written so the quantity on the right is scale invariant. It is the ratio of two energies, the volume binding energy and $\frac{3}{5}$ the Fermi energy. The value of the ratio is the same if scaled variables are used or if λ and $E_F = p_F^2/2M$ are expressed in MeV.

If r_c and g^2/μ^2 are given, one would first calculate a, using the scaling laws. The solution of (39) gives p_i, substitution of this in the scaling law (16) gives the unscaled p_F, from which one can calculate E_F and, from (7), ρ, the density of the nuclear interior. Equation (40) then gives the ratio of the volume binding energy to E_F. On the other hand, if we wish to find the values of r_c and g^2/μ^2 which would result in a preassigned (unscaled) ρ and λ we would calculate p_F, from (7), and E_F. The right side of (40) is then known and the equation can be solved for y_i. Equation (27) gives ap_i, which immediately determines r_c (remember that ap is scale invariant). Equation (39) gives p_i, and the scaling equation (16) gives g^2/μ^2.

IV. AN EXAMPLE

As an example let us take the case $p_i = 5.37$, a value whose relevance will be seen later. From (39) we find $y_i = 1.886$, whence $ap_i = 0.47$, $a = 0.0875$. Equation (40) gives $\lambda / \tfrac{3}{5} p_i^2 = 0.917$, whence $\lambda = 15.86$. If we write (36)

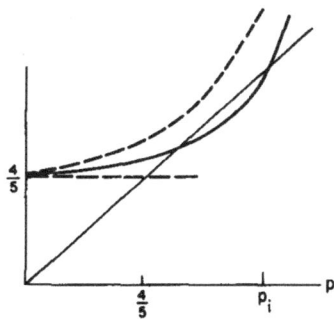

FIG. 4. Graphical representation of Eq. (39).

$$E = -\lambda A + \gamma S, \tag{41}$$

a simple numerical integration gives for the surface tension $\gamma = 2954$. If, following (19), we write

$$\gamma = \lambda p_i^2 T / 2, \tag{42}$$

we find $\frac{1}{2}T = 1.20$. It should be emphasized that the surface energy is the quantity calculated; (19) or (42) is no more than a definition of T; one convenient because, by exhibiting the dimensionality of the surface term, it facilitates scale changes. Thus, in terms of unscaled quantities, the surface energy is

$$E_{\text{sur}} = \frac{1.20}{\mu} \lambda \rho S .$$

If the nuclear density is taken to be $\rho = 3/4\pi r_0^3$, and the area $S = 4\pi R^2$,

$$E_{\text{sur}} = \frac{3 \times 1.20}{\mu r_0} \left(\frac{R}{r_0} \right)^2 \lambda .$$

In terms of γ, we can write the surface energy

$$E_{\text{sur}} = \frac{3\gamma}{p_i^2} \frac{1}{\mu r_0} \left(\frac{R}{r_0} \right)^2 . \tag{43}$$

In this form, since γ and p_i are dimensionless, E_{sur} is also, but its scale factor is now the same as λ (or $g\phi$), that is, the scale given by (17).

Numerical integration of (34) gives the relationship between x and p, and thus ϕ and ρ in terms of x. The Fermi-Thomas well of Fig. 2 is the result of the calculation for this case. Figure 5 shows ρ as a function of x.

V. COULOMB FIELD

The next step in the construction of our model is to include the Coulomb field. We must now distinguish between protons and neutrons, which we do by adding as a variable $t = t(r)$, the local value of the z component of isotopic spin. In terms of t, the proton and neutron densities are $\rho_P = (\frac{1}{2} + t)\rho$, $\rho_N = (\frac{1}{2} - t)\rho$. The Coulomb energy, which must be added to (8), is

$$E_{\text{Coul}} = \frac{1}{2} e \int (\tfrac{1}{2} + t)\rho \Phi \, d\vec{r} \tag{44}$$

with the electrostatic potential Φ given by

$$\Phi(r) = \frac{e}{4\pi} \int \frac{[\frac{1}{2} + t(r')]\rho(r')}{|\vec{r} - \vec{r}'|} \, d\vec{r}' . \tag{45}$$

The mean kinetic energy of a particle, $(3/5)p^2/(1 - ap)^2$ for $t = 0$, also has a t^2 dependence, which we fix by noting that we also know the mean kinetic energy for $t^2 = \frac{1}{4}$, that is, for a hard-sphere Fermi gas of all protons or all neutrons. For only one type of particle the Fermi momentum is increased

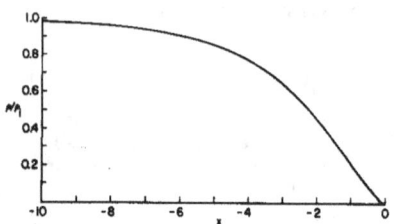

FIG. 5. The density near the surface for the example of Sec. IV.

to $2^{1/3}p$ [p still being determined by (7)]. Also the a term in the denominator is altered because of an effect of the exclusion principle, which prevents a nucleon from approaching another of the same s_z and t_z closer than a distance $1/p_F$. Thus a very short-range interaction is reduced by a factor $\frac{3}{4}$ in a mixture of equal numbers of protons and neutrons (this factor is already included in the coefficient $a = (5/3\pi)r_c$), but by a factor $\frac{1}{2}$ for all neutrons or all protons. For a gas of one kind of particles the coefficient a is therefore reduced to $\frac{2}{3}a$. For $t^2 = \frac{1}{4}$, the mean kinetic energy is $(3/5)2^{2/3}p^2/[1 - (2/3)2^{1/3}ap]^2$. If we write for the mean kinetic energy

$$\frac{3}{5} \frac{p^2(1 + \frac{1}{2}\alpha t^2)}{[1 - ap(1 + \frac{1}{2}\epsilon t^2)]^2} ,$$

with

$$\alpha = 8(2^{2/3} - 1),$$

$$\epsilon = 8(\tfrac{2}{3}2^{1/3} - 1),$$

we obtain the proper limiting values. For small t^2, expanding the denominator gives

$$\frac{3}{5} \frac{p^2}{(1 - ap)^2} + \frac{3}{10} p^2 t^2 \left[\frac{\alpha}{(1 - ap)^2} + 2\epsilon \frac{ap}{(1 - ap)^3} \right] . \tag{46}$$

For $ap < 0.47$, the value of ap_i in Sec. IV, the factor in the square brackets can be approximated to within 2% by a straight line, $\alpha + 8.57ap$. To save writing, we call this $\alpha + \beta p$, with

$$\alpha = 8(2^{2/3} - 1) = 4.7, \quad \beta = 8.57a = 0.75 \tag{47}$$

for $a = 0.0875$. We will therefore use as the expression for the mean kinetic energy

$$\frac{3}{5} \frac{p^2}{(1 - ap)^2} + \frac{3}{10} p^2 (\alpha + \beta p) t^2 . \tag{48}$$

In the nuclear interior t^2 is small, but in the surface layer, for small p, t reaches its limit, $t = -\frac{1}{2}$. In this limit (48) becomes

$$\tfrac{3}{5} p^2 + \tfrac{3}{10} \alpha p^2 \tfrac{1}{4} = \tfrac{3}{5} 2^{2/3} p^2 ,$$

which is the correct value. Thus (48) is designed to be a reasonable representation of the mean kinetic energy over the entire range of the variables.

VI. VARIATION PROBLEM

Since the z component of isotopic spin, $T_z = Z - \frac{1}{2}A$, has been introduced as a variable, the binding energy E is now a function of two variables, A and T_z. Our problem is to minimize E, subject to the conditions $A = \int \rho \, d\vec{r}$ and $T_z = \int t\rho \, d\vec{r}$ constant. We need a second Lagrange multiplier, and the variation problem takes the form

$$\delta(E + \lambda A - \xi T_z) = 0 . \tag{49}$$

It follows from this that

$$\frac{\partial E}{\partial A} = -\lambda , \qquad \frac{\partial E}{\partial T_z} = \xi . \tag{50}$$

If E were written in terms of numbers of protons and neutrons, Z and N, rather than A and T_z, we should have

$$\frac{\partial E}{\partial Z} = -\lambda + \frac{1}{2}\xi , \qquad \frac{\partial E}{\partial N} = -\lambda - \frac{1}{2}\xi . \tag{51}$$

If we wish to find the minimum in the mass-surface valley, that is, for fixed A the value of T_z (or Z) which gives minimum mass, the quantity to be minimized is

$$M = E + \frac{1}{2}(M_N + M_H)A - (M_N - M_H)T_z .$$

The minimum with respect to T_z comes for

$$\xi = M_N - M_H = 0.782 \text{ MeV} . \tag{52}$$

Let us scale $e\Phi$ in the same way as $g\phi$, that is $e\Phi$ and Φ_s are related by the same factor as in (17). Then in scaled form

$$E + \lambda A - \xi T_z = \int \left[\tfrac{1}{2}(\text{grad}\phi)^2 - V\right] d\vec{r} , \tag{53}$$

and now, in place of (22), we have

$$-V = \frac{3}{5} \frac{p^5}{(1 - ap)^2} + \frac{3}{10} p^5(\alpha + \beta p)t^2 + \frac{1}{2}\phi^2 - \phi p^3$$
$$+ \lambda p^3 + \frac{1}{2}(\tfrac{1}{2} + t)p^3\Phi - \xi t p^3 . \tag{54}$$

The variational equations are

$$\Delta\phi = -\partial V/\partial\phi = \phi - p^3 , \tag{55}$$

and $\partial V/\partial p = 0$, $\partial V/\partial t = 0$. To these must be added Laplace's equation for Φ, which in its unscaled form is

$$\Delta\Phi = -e(\tfrac{1}{2} + t)\rho . \tag{56}$$

The equation $\partial V/\partial p = 0$ gives[6]

$$\tfrac{3}{5}p^2 y^2(1 + \tfrac{2}{3}y) + \tfrac{1}{2}p^3(\alpha + \tfrac{4}{5}\beta p)t^2 - \phi + \lambda$$
$$+ (\tfrac{1}{2} + t)\Phi - \xi t = 0 , \tag{57}$$

while $\partial V/\partial t = 0$ gives

$$\tfrac{3}{5}p^2(\alpha + \beta p)t + \Phi - \xi = 0 \tag{58}$$

or

$$t = -\frac{\Phi - \xi}{\tfrac{3}{5}p^2(\alpha + \beta p)} . \tag{59}$$

Multiplying (58) by t and subtracting from (57) gives

$$\tfrac{3}{5}p^2 y^2(1 + \tfrac{2}{3}y) - \tfrac{1}{10}\alpha p^2 t^2 + \tfrac{1}{2}\Phi - \phi = -\lambda \tag{60}$$

as the generalization of (28).

In a similar way, we can use (58) to eliminate the $tp^3\Phi$ term from (54) and obtain

$$-V = (\tfrac{3}{5}p^2 y^2 - \phi + \lambda + \tfrac{1}{4}\Phi - \tfrac{1}{2}\xi t)p^3 + \tfrac{1}{2}\phi^2 . \tag{61}$$

VII. NUCLEAR INTERIOR

We again divide the nucleus into a surface layer and an interior. In carrying out the solution of the equations it is important to take a self-consistent view, that is, to take seriously the stricture $R \gg 1/\mu$. In the interior the variables ϕ, p, and t are changing because Φ is a function of r; however, the changes are adiabatic, that is, the variables change little in a distance $1/\mu$; their scale of change is R rather than $1/\mu$. Thus, in (5), grad$\phi \sim \phi/R$ and the $(\text{grad}\phi)^2$ term is of order $1/(\mu R)^2$ compared with the $\mu^2\phi^2$ term, and can be neglected. The variational equation, (55), then becomes $\partial V/\partial\phi = 0$, or $\phi = p^3$, which holds throughout the interior.

The boundary condition which must be satisfied at the nuclear surface, $r = R$, is again that the pressure be zero, i.e., $V = 0$, at $r = R$. We shall use the subscript R to denote the values of variables at $r = R$. If we put $\phi = p^3$ in (61), we find that $V = 0$ gives the condition

$$\tfrac{3}{5}p_R^2 y_R^2 - \tfrac{1}{2}p_R^3 + \tfrac{1}{4}\Phi_R - \tfrac{1}{2}\xi t_R = -\lambda . \tag{62}$$

If we eliminate λ by subtracting (62) from (60) with $\phi = p_R^3$, we find the generalization of (39), which now reads

$$p_R = \tfrac{4}{5}y_R^3 - \tfrac{1}{5}\alpha t_R^2 + (\tfrac{1}{2}\Phi_R + \xi t_R)/p_R^2 . \tag{63}$$

Using (63), (62) can be written

$$\tfrac{2}{3}y_R^3 - y_R^2 = \frac{\lambda - \xi t_R}{\tfrac{3}{5}p_R^2} + \tfrac{1}{6}\alpha t_R^2 , \tag{64}$$

the generalization of (40).

In the nuclear interior, $r \leq R$, Eq. (60), with $\phi = p^3$ and t expressed explicitly by (59), is a quadratic equation for Φ, whose solution gives us Φ as a function of p. The electrostatic potential also satisfies (56), which, in terms of the scaled variables Φ and ρ, reads

$$\Delta\Phi = -\frac{\mu^2}{g} e^2(\tfrac{1}{2}+t)p^3. \qquad (65)$$

If we introduce a dimensionless length for the nuclear interior,

$$r = \left(\frac{g^2}{\mu^2}\frac{1}{e^2}\frac{1}{p_0^{\,3}}\right)^{1/2} r_s \qquad (66)$$

(65) takes the form

$$\Delta\Phi = -(\tfrac{1}{2}+t)p^3/p_0^{\,3}. \qquad (67)$$

A standard density, $\rho_0 = p_0^{\,3}$, has been introduced into (66) and (67) as a matter of later convenience. The reason for using (66) rather than $r = r_s/\mu$, for the nuclear interior is that the scaling length it gives remains finite in the adiabatic limit, $\mu \to \infty$ with g^2/μ^2 remaining constant. Thus, even including the Coulomb term, the properties of the nuclear interior depend only on g^2/μ^2. The solution of (67) gives Φ as a function of r, and, since Φ is known in terms of p from (60), gives p, hence ρ, t, and ϕ, as functions of r.

VIII. CHOICE OF g^2/μ^2 AND a

The property of nuclear matter most directly related to the magnitude of the repulsive core radius is its compressibility. We shall accordingly try to choose a to fit the evidence on nuclear compressibility which appears in the results Hofstadter and his group[7] obtained in their experiments on the scattering of high energy electrons by nuclei. With the nuclear radius written $R = r_0 A^{1/3}$, the results are shown in Table I. We see that the nucleus expands under the influence of the repulsive Coulomb force, r_0 increasing from 1.06 fm at $Z = 23$, to 1.075 fm at $Z = 50$, to 1.096 fm at $Z = 79$. As the higher accuracy of the figure listed indicates, gold was the element studied most thoroughly. In view of these figures, we adopted $r_0 = 1.05$ fm for the limit of no Coulomb interaction ($e = 0$).[8] With two rather arbitrarily chosen values of a we then went through the procedure described in the next section to determine the nuclear radius for Au, and in the light of the results, guessed the value $a = 0.0875$. The calculation for this value led to the result $r_0 = 1.093$ fm for Au, which seemed close enough.

With $a = 0.0875$, Eq. (39) gives the values of p_i, ap_i, and y_i quoted in Sec. IV. We shall use a subscript 0 to denote the value of quantities, including r_0, in the $e = 0$ limit; thus $p_0 = 5.37$, $ap_0 = 0.47$. For $r_0 = 1.05$ fm, the unscaled nuclear density is $\rho_0 = 3/4\pi r_0^{\,3}$, Eq. (7) gives $p_{F0} = (9\pi/8)^{1/3}/$

TABLE I. Electron scattering (Hofstadter *et al.*): $R = r_0 A^{1/3}$.

	r_0 (fm)	
$_{20}^{40}$Ca	1.06	
$_{23}^{51}$V	1.07	1.06
$_{27}^{59}$Co	1.05	
$_{49}^{115}$In	1.08	1.075
$_{50}^{122}$Sn	1.07	
$_{79}^{197}$Au	1.096	1.096
$_{83}^{209}$Bi	1.09	

r_0, and (16) gives

$$\frac{g^2}{\mu^2} = \frac{\pi(3\pi^2)^{1/3}}{2}\frac{r_0}{M}p_0. \qquad (68)$$

The Fermi energy $E_{F0} = p_{F0}^{\,2}/2M = 43.63$ MeV. In scaled units it is $p_0^{\,2} = 28.84$. The ratio

$$p_0^{\,2}/E_{F0} = 0.661 \qquad (69)$$

gives the conversion factor for changing energies in MeV to scaled energies. The scale factor for length in the nuclear interior is

$$\left(\frac{g^2}{\mu^2 e^2 p_0^{\,3}}\right)^{1/2} = \left[\frac{(3\pi^2)^{1/3}}{8}\frac{1}{\alpha}\frac{r_0}{M}\frac{1}{p_0^{\,2}}\right]^{1/2} = 0.637 \text{ fm}, \qquad (70)$$

where $\alpha = e^2/4\pi$ is the fine structure constant. Since ap is scale invariant, we have $(5/3\pi)r_c p_{F0} = ap_0$, or

$$r_c = \frac{3\pi}{5}\times 0.47\left(\frac{8}{9\pi}\right)^{1/3} r_0 = 0.61 \text{ fm}.$$

The value found in nucleon-nucleon scattering is $r_c = 0.5$ fm.

IX. SOLUTION IN THE NUCLEAR INTERIOR

The procedure in solving the problem for the nuclear interior is to first choose a value of Φ_R, guided by the unscaled relation

$$e\Phi_R = \frac{e^2}{4\pi}\frac{Z}{R}. \qquad (71)$$

In our further work we shall always be discussing the most stable nucleus for a given A, so by (52), we shall take

$$\xi = 0.782 \text{ MeV} = 0.517 \qquad (72)$$

in scaled units. Equation (63), with t_R given by (59), can then be solved for p_R, and (64) gives λ.

Equation (60) can now be solved for Φ as a function of p, and, inverting this relationship,

$\rho_p/\rho_0 = (\frac{1}{2} + t)\rho/\rho_0$

can be determined as a function of Φ. The relationship, in the case of Au, is shown in Fig. 6, where the value of Φ_R is indicated. If the nucleus had a uniform charge density, the value of Φ at the center, $r = 0$, would be $\Phi(0) = \frac{3}{2}\Phi_R$. Because of the Coulomb repulsion, the protons will be driven towards the outside and $\Phi(0)$ will be somewhat less than $\frac{3}{2}\Phi_R$ [for a surface charge $\Phi(0) = \Phi_R$]. In the range between Φ_R and $\frac{3}{2}\Phi_R$ the charge density is very nearly a linear function of Φ; it can be represented by

$$\frac{\rho_P}{\rho_0} = b - K^2\Phi \qquad (73)$$

with an error nowhere more than 0.7%. Equation (67) then takes the form

$$\Delta\Phi - K^2\Phi = -b. \qquad (74)$$

The solution of (74) satisfying at $r = R$ the boundary condition

$$(R/\Phi)d\Phi/dR = -1,$$

which follows immediately from Gauss's theorem and (71), is

$$\Phi(r) = \frac{b}{K^2}\left(1 - \frac{\sinh Kr}{Kr\cosh KR}\right). \qquad (75)$$

For $r = R$, Eq. (75) gives

$$1 - \frac{\tanh KR}{KR} = \frac{K^2}{b}\Phi_R, \qquad (76)$$

and for $r = 0$, it gives

$$\Phi(0) = \frac{b}{K^2}\left(1 - \frac{1}{\cosh KR}\right). \qquad (77)$$

With the constants b and K^2 determined by the process we have described, (76) gives us the scaled nuclear radius, and multiplication by the scale factor, Eq. (70), gives us the radius in femtometers. When we start by assuming a value of $e\Phi_R$ we do not know exactly what value of Z will come out of the calculation, since (71) fixes only the ratio Z/R. Now, knowing R, Eq. (71) gives us the value of Z.

Equation (75) gives Φ as a function of r, and since the relationship between Φ and ρ is known, all quantities $(\phi, \rho, \rho_P$, the neutron density ρ_N, and $t)$ are known as functions of r. At the start A is not known either; we can now determine it by doing the integral in Eq. (9). The scaling problem can be circumvented by calculating the ratio of the mean density to ρ_0,

$$\langle\rho/\rho_0\rangle = \int (\rho/\rho_0)d\vec{r}\,\frac{3}{4\pi R^3},$$

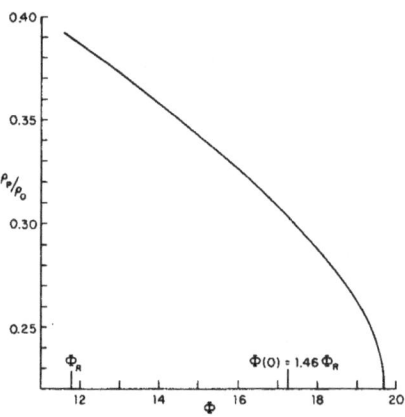

FIG. 6. The charge density as a function of Φ for Au.

which is scale invariant, and writing

$$A = \frac{4\pi}{3}R^3\rho_0\langle\rho/\rho_0\rangle = \left(\frac{R}{r_0}\right)^3\langle\rho/\rho_0\rangle. \qquad (78)$$

For the case of Au, the value $e\Phi_R = 17.85$ MeV was chosen, or, in scaled units using (69), $\Phi_R = 11.80$. Equation (63) gives $p_r = 5.236$, (64) gives $\lambda = 12.78$. The constants appearing in (73) are found to be $b = 0.579$, $K^2 = 0.01588$. Equation (76) gives $R = 9.97 = 6.35$ fm on using the scaling factor, Eq. (70), and (71) gives $Z = 78.7$. Equation (77) gives $\Phi(0) = 17.27 = 1.46\Phi_R$, the corresponding value of p is $p = 5.026$.

Numerical integration gives $\langle\rho/\rho_0\rangle = 0.888$, and, from (78), $A = 196.4$. For this value of A the empirical mass formula, in the form quoted by Bohr and Mottelson,[2] Vol. 1, p. 169, gives for the Z of minimum mass $Z = 78.4$, which is to be compared with the value we obtained, $Z = 78.7$. Our model thus predicts the value of Z/A to within $\frac{1}{2}$%.

X. BINDING ENERGY

In place of (33), we now have for the energy

$$E = -\lambda A + \xi T_Z + \int_0^R (-V)d\vec{r} + S\int_0^{*R}(-2V)^{1/2}d\phi. \qquad (79)$$

In the volume term $-V$ is given by (61) with $\phi = p^3$.

The interpretation of (79) is this: $-\lambda + \xi\langle t\rangle$ represents the binding energy of a particle near the nuclear surface, and the volume term corrects for the fact that nucleons in the interior are less firmly bound than those near the surface. The surface term is the same as in the discussion of Sec. II except that (60) and (61) replace (28) and (22). It is implicit in our separation of surface and volume effects that the thickness of the surface layer is small compared to R. Accordingly Φ, whose scale of variation is set by R, may be treated as a constant, $\Phi = \Phi_R$, in the surface layer. As p decreases in passing through the surface layer we reach a value ($p = 2.41$ in the case of Au) at which $t = -\frac{1}{2}$. Beyond this the surface layer consists only of neutrons, and in place of (61) one uses (54) with $t = -\frac{1}{2}$.

In our present notation, Eq. (43) takes the form

$$E_{sur} = \frac{3\gamma}{\rho_0^3} \frac{1}{\mu r_0} \left(\frac{R}{r_0}\right)^3. \tag{80}$$

The volume energy can be written

$$E_{vol} = \left[-\lambda + \xi(t) + \int_0^R (-V) d\bar{\tau}/A \right] A. \tag{81}$$

XI. DETERMINATION OF μ

The parameter μ can be determined by fitting the binding energy, given by the sum of (80) and (81), to the value given by the empirical mass formula.

For Au, where $\lambda = 12.78$, a numerical integration gives $\int_0^R (-V) dr/A = 0.72$. The surface tension is found to be $\gamma = 2951$, virtually unchanged from that found in Sec. IV. The mass formula gives $E/A = -7.95$ MeV $= -5.25$ on using the scaling factor (69). Using the values previously given, we have

$$\frac{E}{A} = -12.10 + \frac{10.65}{\mu r_0} = -5.25,$$

which gives

$$\begin{aligned} 1/\mu r_0 &= 0.643, \\ 1/\mu &= 0.68 \text{ fm}, \end{aligned} \tag{82}$$

or

$$\mu = 290 \text{ MeV}.$$

Equation (68) then gives

$$g^2/4\pi = 1.00.$$

The mass we find lies near two π masses, the threshold for two π exchange.

XII. DENSITY DISTRIBUTIONS

The particle density in the nuclear interior and in the surface layer have been calculated essentially independently; to fit them together one must determine the position of the center of gravity of the surface layer, call it x_{CG}, and place this point at R. By calculating the integral through the surface layer

$$\lim_{r \to \infty} \int_R^0 \rho dx = -\rho_R (x - x_{CG})$$

we find $x_{CG} = -2.66 = -1.81$ fm, on using the scaling factor (83). Because of the large value found for $1/\mu$, the thickness of the surface layer $T = 2.4/\mu = 1.7$ fm is in fact not very small compared to the nuclear radius, $T/R = 0.26$. Since the adiabatic approximation is imperfectly satisfied, we have a problem, for example for the density ρ, in fitting the surface and interior densities together: By the time the surface density has approached its limiting value ρ_R, the interior density has dropped appreciably from its value, also ρ_R, at $r = R$. To make the two density curves fit smoothly we have, at each interior point, $r < R$, multiplied the density given for the surface layer by $\rho(r)/\rho_R$ as given by the interior solution. The same procedure was followed for the other quantities, e.g., for ϕ in constructing the Fermi-Thomas wells.

Figure 7 show the Fermi-Thomas wells for neutrons and protons in Au, and Fig. 8 shows the neutron, proton, and total densities. An interesting point is that the outermost shell of the nucleus, of thickness 0.4 fm, consists solely of neutrons.

Before comparing the calculated proton distribution with measured ones, the question comes up of the effect of barrier penetration on the distribution. In the Fermi-Thomas treatment the density is taken classically, in that it is put at zero, for a particle of given energy in the well, beyond the classical turning point. Figure 9 shows the change in the proton distribution resulting from using instead a WKB approximation.

We chose Au as the heavy element for which to make our calculations because that was the element Hofstadter and his collaborators studied most carefully in their earlier work. However, we have since discovered that more recent data and calculations[9] exist for $^{82}Pb^{208}$. Our calculated charge distribution for Pb is shown in Fig. 10 together with the distribution deduced from the experiments. The agreement is good except for the distinct difference in shape near the shoulder. To fit the observed distribution in this region would require a thinner surface layer (a meson mass of about 500 MeV), but then the curves would disagree in the low density region, for large r. A model which gives $\mu = 500$ MeV has been discussed by the author,[10] it is slightly more complicated than the present model in that it requires an additional type of force, an exchange force between nucleons.

XIII. LIMITING CASE, Z = 121

The behavior of the curve in Fig. 6 raises a question: What happens if Φ_R is so large that $\Phi(0)$,

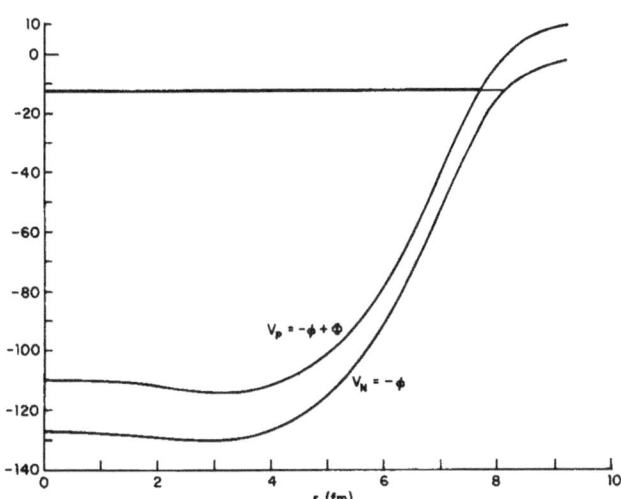

FIG. 7. The Fermi-Thomas wells for neutrons and protons for Au.

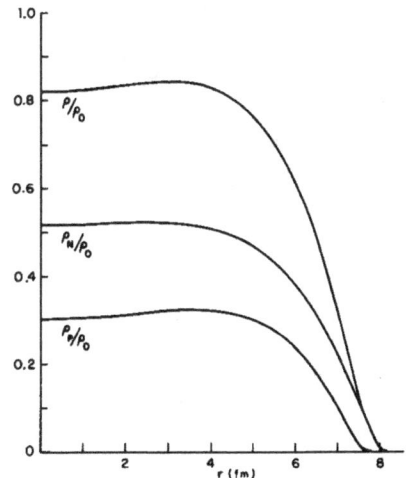

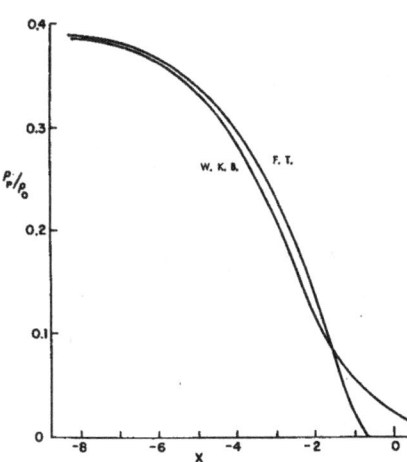

FIG. 8. The neutron, proton, and total densities for Au.

FIG. 9. The proton density corrected by the WKB approximation.

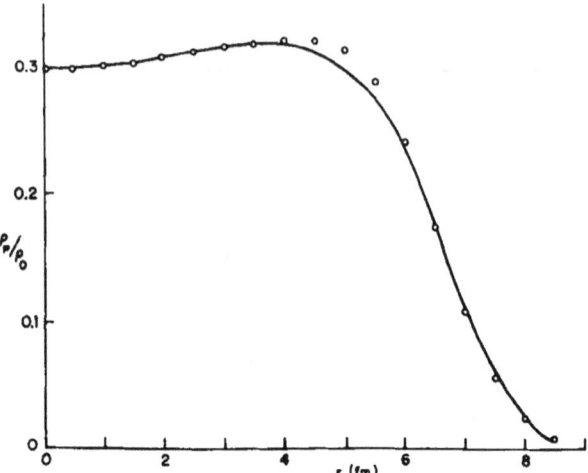

FIG. 10. The charge distribution calculated for ^{208}Pb. The circles give the charge density deduced from the measurements of Hofstadter *et al.*

which is a little less than $\frac{2}{3}\Phi_R$, is beyond the value of Φ at which the curve reverses, which it does for $\rho_p/\rho_0 < 0.22$? The answer follows from the significance of the curve: It represents the relationship between ρ_p and Φ required for dynamical equilibrium in the interior of the nucleus. If the density falls too far the attractive forces between nucleons are not enough to balance the Fermi pressure. The reversal of the curve means that it would then require an electric field, which it does for $\rho_p/\rho_0 < 0.22$? The reversal of the curve means that it would then require an electric field, which it does in the opposite direction (inwards) to restore equilibrium. Since, in fact, the direction of the electric field cannot reverse, what happens is that the center of the nucleus is emptied, and an inner surface layer forms at the radius at which Φ reaches the maximum permitted by the curve of Fig. 6.

We can ask for the limiting Z beyond which a central hole begins to develop, that is, for the Φ_R such that $\Phi(0)$ is the limit permitted by the ρ_p/ρ_0 curve. This problem is a little more difficult than the calculation for Au, since the linear approximation of (73) cannot be used, and Laplace's equation must be integrated numerically. The answer is $\Phi_R = 15.06$, $\Phi(0) = 21.62 = 1.44\Phi_R$, $R = 7.67$ fm, $Z = 121.4$, $A = 330$. The Z/A ratio is exactly that given by the empirical mass relation.

XIV. COMMENT

There have been many papers published on Fermi-Thomas nuclear models, some of them quite successful and dealing much more realistically and completely with nuclear properties than the present paper. An extensive bibliography is given in an article by Myers and Swiatecki,[11] and discussion of some of the work in an article by Bethe.[12] None of the effects discussed in this paper is novel; all have been elucidated in earlier works. The aim of the present paper has been simplicity, rather than realism, in the hope that a simpler treatment may provide more physical insight for a nonexpert reader. The treatment differs from earlier ones in its description of the effect of the repulsive core in terms of the results of the theory of a hard-core Fermi gas in the form given by Bohr and Mottelson.[2] Such a treatment may possibly be of some value even in a more sophisticated theory. Another novelty is the use, following Lee and Wick,[1] of the meson potential as a variable in the variation problem, a choice both physically meaningful and leading to a felicitous mathematical formulation.

*This research was supported in part by the U.S. Energy Research and Development Administration.

[1] T. D. Lee and G. C. Wick, Phys. Rev. D 9, 2291 (1974); T. D. Lee, Rev. Mod. Phys. 47, 267 (1975).

[2] A. Bohr and B. R. Mottelson, *Nuclear Structure* (Benjamin, New York, 1969), Vol. 1, p. 256.

[3] Note that the *total* energy E scales differently. However, the energy per particle, $E/\int \rho d\vec{r}$, also scales according to (17).

[4] Note that in virtue of (25), there is no difference between $dV/d\phi$ and $\partial V/\partial\phi$.

[5] For λ equal to zero the curve of Fig. 3 would be just the negative of the dotted curve of Fig. 1, though with a different abscissa scale. Because of the arbitrary insertion of a minus sign in defining V, for the purpose of making the dynamical analogy, stability and instability are reversed in the two cases. Also note that, according to (24), $\partial V/\partial\phi = 0$ means $\phi = \rho^3$, the condition (10), used to derive (12).

[6] In carrying out the variation it must be remembered that ϕ is also a function of ρ and t. The effect of varying ϕ, as can most easily be seen by writing (44) as a double integral using (45), is to double the contribution the Coulomb term would give if ϕ were not varied, i.e., the factor $\frac{1}{2}$ in the Coulomb term is removed.

[7] B. Hahn, D. G. Ravenhall, and R. Hofstadter, Phys. Rev. 101, 1131 (1956).

[8] I am indebted to Dr. W. J. Swiatecki for pointing out to me a logical flaw in this argument. For nuclei as light as Ca the thickness of the surface layer is not very small compared to the radius, and T/R corrections may also be of significance. In consequence, the choice $r_0 = 1.05$ fm must be viewed as rather arbitrary.

[9] C. W. DeJager, H. DeVries, and C. DeVries, At. Data Nucl. Data Tables 14, 501 (1974); B. Dreher, J. Friedrich, K. Merle, H. Rothhaas, and G. Lührs, Nucl. Phys. A235, 219 (1974); J. Heisenberg, R. Hofstadter, J. S. McCarthy, I. Sick, B. C. Clark, R. Herman, and D. G. Ravenhall, Phys. Rev. Lett. 23, 1402 (1969).

[10] R. Serber, in *High-Energy Physics and Nuclear Structure—1975*, edited by D. E. Nagle *et al.* (American Institute of Physics, New York, 1975), p. 700.

[11] W. D. Myers and W. J. Swiatecki, Ann. Phys. (N.Y.) 55, 395 (1969).

[12] H. A. Bethe, Phys. Rev. 167, 879 (1968).

Reprinted from THE PHYSICAL REVIEW, Vol. 72, No. 11, 1008–1016, December 1, 1947

The Production of High Energy Neutrons by Stripping*

R. SERBER

Radiation Laboratory, Department of Physics, University of California, Berkeley, California

(Received August 28, 1947)

When a target is bombarded with high energy deuterons, a narrow beam of high energy neutrons is produced by a process in which the proton in the deuteron strikes the edge of the nucleus and is stripped off, while the neutron misses and continues on its way. The cross section for this stripping process is $\sigma = \frac{1}{2}\pi R R_d$, where R is the nuclear radius and R_d is the deuteron radius, or $\sigma = 5 A^{\frac{1}{3}} \times 10^{-26}$ cm^2. The yield of neutrons from a $\frac{1}{2}$-in. Be target (in which the energy loss for 190-Mev deuterons is 20 Mev) is nearly 2 percent. The neutrons come out with an energy spread around $\frac{1}{2}E_d$ having a half-width $\Delta E_{\frac{1}{2}} = 1.5(E_d \epsilon_d)^{\frac{1}{2}}$. Here E_d is the kinetic energy of the deuteron, ϵ_d its binding energy.

For light nuclei the half-width of the neutron angular distribution is $\Delta \theta_{\frac{1}{2}} = 1.6(\epsilon_d/E_d)^{\frac{1}{2}}$. The half-width increases somewhat with atomic number, primarily because of the deflection of the deuteron by the Coulomb field as it approaches the nucleus, and, to a lesser extent, because of multiple scattering in the target. The increase in half-width from Be to U is about 25 percent. The calculated half-widths and angular distributions agree well with the measurements of Helmholz, McMillan, and Sewell.

An equal number of high energy protons are produced by stripping processes in which it is the neutron that hits the nucleus.

I. INTRODUCTION

THERE are several processes by which high energy neutrons may be produced when a target is bombarded by high energy deuterons. A deuteron passing at some distance from an atomic nucleus, say two or three times the nuclear radius, may be disintegrated by the Coulomb field of the nucleus.[1] Or, when the deuteron grazes the edge of the nucleus, the proton may strike it and be stripped off, while the neutron misses and continues with almost the velocity of the incident deuteron. And, finally, a high energy neutron can be produced by a direct collision between one of the particles of the deuteron and a nuclear particle.

It is the second process, the stripping process, which will be discussed in this paper. Its characteristics depend primarily on the fact that the deuteron is a very loosely-bound system, the proton and neutron actually spending most of their time outside the range of their mutual forces. For deuterons of kinetic energy considerably larger than the deuteron binding energy, the collision time of the proton with a nuclear particle will be small compared to the period of the relative motion of neutron and proton within the deuteron, and the momentum transferred to the proton will be large compared to the mo-

mentum of the relative motion. The proton is thus effectively stripped off instantaneously; there is no reaction on the neutron, which continues its flight with the momentum it had at the instant of collision. This momentum is the sum of the momentum attributable to the motion of the center of mass of the deuteron, plus that attributable to the motion of the neutron within the deuteron. The former is $p_0 = (ME_d)^{\frac{1}{2}}$, where M is the proton mass and E_d is the kinetic energy of the deuteron, while the latter is of the order $p_1 = (M\epsilon_d)^{\frac{1}{2}}$, with $\epsilon_d = 2.18$ Mev the binding energy of the deuteron. The neutron will therefore emerge within an angle to the direction of the deuteron beam of about $\theta \sim (p_1/p_0) = (\epsilon_d/E_d)^{\frac{1}{2}} \sim 6°$ for $E_d = 190$ Mev. The energy of the neutrons will mostly be in a band given by

$$E = (p_0 \pm p_1)^2/2M \sim \frac{1}{2}E_d[1 \pm 2(\epsilon_d/E_d)^{\frac{1}{2}}]$$
$$\sim \frac{1}{2}E_d \pm 20 \text{ Mev.}$$

The most striking feature of the stripping effect is thus the production of a very narrow cone of neutrons with energy about half that of the deuterons. This prediction has been confirmed on the 184-in. cyclotron in a series of experiments carried out by Helmholz, McMillan, and Sewell; these are reported in the preceding paper.

It is just the narrowness of the cone which distinguishes the stripped neutrons from those produced by direct nuclear encounters, although

---

* Originally reported at the July 1947 meeting of the American Physical Society.
[1] J. R. Oppenheimer, Phys. Rev. **47**, 845 (1935); S. M. Dancoff, Phys. Rev. **72**, 163 (1947).

the total number of neutrons from the two processes are expected to be about the same. For the latter, the formula for the characteristic angle would have, in place of ϵ_d, an energy ϵ_n of the order of the kinetic energy of particles in the nucleus, $\epsilon_n \sim 20$ Mev. The cone would be over three times as wide, and, since the intensity in the forward direction is proportional to the inverse square of the cone width, only a 10 percent contribution might be expected from these neutrons. Furthermore, it can be shown that in consequence of the fact that the struck particle is bound in the nucleus, collisions with small momentum transfers are discouraged, and the forward intensity is smaller even than the above estimate would indicate. The other process which produces high energy neutrons, the disintegration by the electric field, has a cross section which is proportional to Z^2, and small compared to the stripping cross section except for the heaviest nuclei. Even for U, according to estimates by Dancoff,[2] the cross section for the electric field disintegration is only one-quarter the stripping cross section.

II. THE STRIPPING CROSS SECTION

If the incident deuterons have high energy, the neutron passes the nucleus so quickly that its displacement perpendicular to the line of motion of the deuteron during the time of passage is negligible. In a typical impact, the proton will fail to clear the edge of the nucleus by a distance of the order of the "deuteron radius", $R_d = \frac{1}{2}\hbar/(M\epsilon_d)^{\frac{1}{2}} = 2.1 \times 10^{-13}$ cm, while the neutron will miss by a like distance. The proton will strike the nucleus a distance $l = (2RR_d)^{\frac{1}{2}}$ in front of a plane through the center of the nucleus. Here R is the nuclear radius, which we suppose appreciably larger than R_d. The neutron traverses this distance in a time l/v, where v is the deuteron velocity. The neutron (or proton) will have a velocity normal to the direction of the deuteron motion of the order $(\epsilon_d/M)^{\frac{1}{2}}$, so its displacement in this direction is $(\epsilon_d/M)^{\frac{1}{2}}l/v$. This displacement is unimportant provided it is small compared to R_d, i.e., $(\epsilon_d/M)^{\frac{1}{2}}l/v < R_d$, a relation which may be rewritten, remembering that $E_d = Mv^2$,

$$E_d > 2(R/R_d)\epsilon_d. \qquad (1)$$

[2] S. M. Dancoff, private communication.

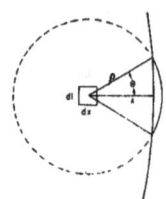

Fig. 1.

The limiting energy given by (1) is, even for the heaviest nuclei, $E_d > 20$ Mev.

The above argument shows that only the (projected) positions of neutron and proton in a plane perpendicular to the deuteron motion need be considered in calculating the cross section: we have only to ask for the probability that at the instant of collision the proton will be within a circle in this plane of radius equal to the nuclear radius, while the neutron will be outside it. Consider a collision in which the separation between proton and neutron (projected in the plane) is ρ. The cross section for the proton hitting a distance x inside the nucleus, within an interval dx, and within an interval dl along the circumference of the nucleus, is just $dxdl$. In the interest of simplicity, we suppose the nuclear radius, R, large compared to the deuteron radius, R_d, so that the curvature of the edge of the nucleus within a distance R_d can be neglected, and the edge considered straight. The probability that the neutron will miss the nucleus is just the fraction of the circumference of a circle of radius ρ which lies outside the nucleus (see Fig. 1); its value is θ/π. The total cross section for proton hitting and neutron missing is thus

$$\sigma(\rho) = \int\int (\theta/\pi)dxdl.$$

The integration over dl gives just the circumference of the nucleus, $2\pi R$. Since $x = \rho\cos\theta$, we have $dx = -\rho\sin\theta d\theta$, and the integral over dx becomes

$$(\rho/\pi)\int_0^{\pi/2} \theta \sin\theta d\theta = \rho/\pi,$$

so

$$\sigma(\rho) = 2R\rho. \qquad (2)$$

Equation (2) gives the cross section when proton and neutron are separated a distance ρ;

R. SERBER   1010

to get the total cross section we must multiply (2) by the probability of finding such a separation, and integrate over all values of ρ. If $\psi_d(r)$ is the wave function of the deuteron in its ground state, $|\psi_d(r)|^2 d\mathbf{r}$ is the probability of finding an $n-p$ separation r in the three-dimensional volume element $d\mathbf{r}$. Thus, introducing cylindrical coordinates, the probability of a separation ρ is

$$2\pi\rho d\rho \int_{-\infty}^{\infty} |\psi_d(r)|^2 dz,$$

and the total stripping cross section is

$$\sigma = 4\pi R \int_{-\infty}^{\infty} dz \int_{0}^{\infty} |\psi_d(r)|^2 \rho^2 d\rho. \quad (3)$$

We can change the variable of integration from z to r by using the relation $r^2 = \rho^2 + z^2$, which gives $dz = rdr/(r^2-\rho^2)^{\frac{1}{2}}$, and transforms (3) into

$$\sigma = 8\pi R \int_{0}^{\infty} |\psi_d(r)|^2 rdr \int_{0}^{r} (\rho^2 d\rho)/(r^2-\rho^2)^{\frac{1}{2}}.$$

The integration over ρ gives $(\pi/4)r^2$, so finally

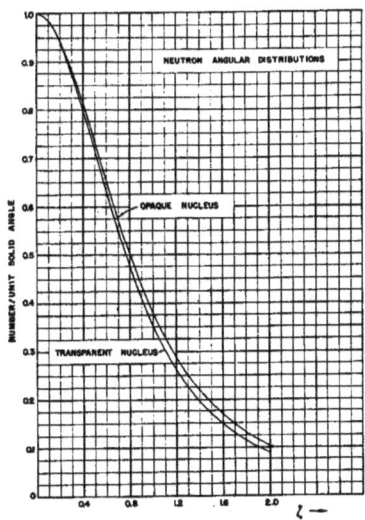

FIG. 2. Neutron angular distributions for transparent and opaque nuclei. Relative number of neutrons per unit solid angle plotted against $\zeta = \theta/\theta_0$.

$$\sigma = 2\pi^2 R \int_{0}^{\infty} |\psi_d(r)|^2 r^2 dr$$

$$= (\pi/2)R \int r |\psi_d(r)|^2 d\mathbf{r}, \quad (4)$$

remembering that $d\mathbf{r} = 4\pi r^2 dr$. The integral in (4) has a simple interpretation: it gives just $\bar{r}$ the average separation of neutron and proton in the deuteron. Calling this separation R_d, we have for the stripping cross section

$$\sigma = (\pi/2)RR_d. \quad (5)$$

If we take

$$\psi_d = (\alpha/2\pi)^{\frac{1}{2}}e^{-\alpha r}/r, \quad \alpha = (M\epsilon_d)^{\frac{1}{2}}/\hbar, \quad (6)$$

we find $R_d = 1/(2\alpha) = 2.1 \times 10^{-13}$ cm, the result previously quoted.

In actuality, the deuteron wave function has the form (6) only outside the range of $n-p$ forces. If the finite range of the forces were taken into account, a somewhat larger value of R_d would be obtained. However, in considering the stripping effect we are only interested in the narrow neutron beam which comes off nearly in the forward direction, and these neutrons are produced in collisions in which the neutron and proton are outside the range of their forces at the instant of collision. Collisions which occur with neutron and proton within the range of the forces will give rise to a wide angular distribution, similar to that resulting from direct nuclear encounters, and may be lumped with the latter effect. Thus, within the limits of unambiguity inherent in the separation of the effects, it is proper to ignore the finite range of the forces.

The derivation of (5) has been carried out as if the nucleus were completely opaque to the neutron and proton. In fact, there will be a finite mean-free path, of the order of 4×10^{-13} cm, for a particle to make a collision in traversing nuclear matter. So in some cases, even though the neutron does not miss the nucleus, it may pass through the edge without being disturbed. However, this effect is balanced out by the approximately equal number of cases in which the proton passes through the edge without a collision.

If we take $R = 1.5A^{\frac{1}{3}} \times 10^{-13}$ cm, (5) becomes

$$\sigma = 5A^{\frac{1}{3}} \times 10^{-26} \text{ cm}^2. \quad (7)$$

The stripping cross section ranges from 0.1 barn

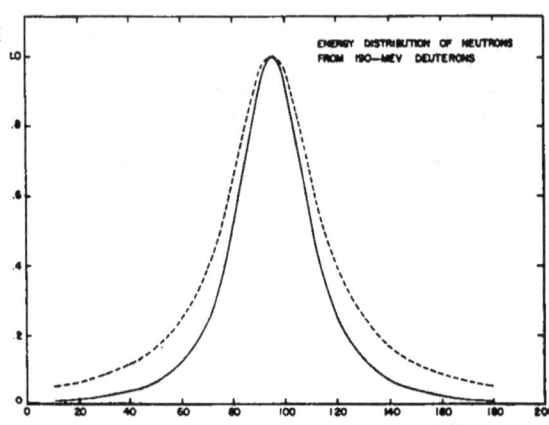

FIG. 3. Energy distribution of neutrons from 190-Mev deuterons. Solid curve, opaque nucleus; dotted curve, transparent nucleus.

for Be to 0.3 barn for U. The yield of neutrons in one passage of the deuteron beam through a $\frac{1}{16}$-in. target, such as is often used in the 184-in. cyclotron, is 1/500 for Be, 1/400 for U.

There is, of course, an equal yield of high energy protons from collisions in which it is the neutron that strikes the nucleus, and the proton which misses.

III. ANGLE AND ENERGY DISTRIBUTIONS

We shall first calculate the angular distribution with the assumption that the nucleus is completely transparent to neutrons. The reason, in addition to the simplicity of this case, is that the model of a transparent nucleus is one limiting case, of which the model of a completely opaque nucleus is the other. Since, as we shall see, the angular distributions to be expected of these two limits turn out to be very little different, we gain by the comparison a considerable confidence in the reliability of the results. Alternatively, we can describe the transparent case as that to be expected in the limit of a very small nucleus, $R \ll R_d$. In our treatment of the opaque case we consider $R \gg R_d$. Treatment of the opposite limit therefore provides some insight into the error likely to be caused by applying the opaque model to light nuclei, where R is not very large compared to R_d, and effects of curvature of the edge of the nucleus and transparency might be expected to show.

The simplicity of the transparent nucleus case lies in the fact that the distribution of neutron momenta due to its motion in the deuteron is just that characteristic of the ground state of the deuteron, without any modification resulting from adding a condition that the neutron has to miss the nucleus.

The probability, $P(\mathbf{p})$, that the neutron in the deuteron has a momentum $\mathbf{p}$ in the interval $d\mathbf{p}$ is

$$P(\mathbf{p}) = |\psi(\mathbf{p})|^2,$$

$$\psi(\mathbf{p}) = (1/h^{\frac{3}{2}}) \int \psi_d \exp[-(i/\hbar)\mathbf{p}\cdot\mathbf{r}]d\mathbf{r}. \quad (8)$$

Using (6), we find

$$\psi(\mathbf{p}) = (1/\pi)[(M\epsilon_d)^{\frac{1}{4}}/(M\epsilon_d + p^2)],$$
$$P(\mathbf{p}) = (1/\pi^2)[(M\epsilon_d)^{\frac{1}{2}}/(M\epsilon_d + p^2)^2]. \quad (9)$$

To get the total momentum of the emergent neutron, we have to add to $\mathbf{p}$ the momentum in the z direction

$$p_0 = (ME_d)^{\frac{1}{2}}[1 + (E_d/8Mc^2)]$$

caused by the motion of the center of mass of the deuteron. The second term in the bracket is a small relativistic correction term.

If we denote by $p_\perp$ the magnitude of the component of $\mathbf{p}$ perpendicular to z, the angle of emergence of the deuteron is

$$\theta = (p_\perp/p_0). \quad (10)$$

R. SERBER

More strictly, we should write $\theta = p_\perp/(p_0+p_z)$. However, p_z is small compared to p_0, and since p_z is equally likely to be positive or negative, the correction terms in the angular distribution linear in p_z/p_0 drop out, and we are left with a correction only of the order $(p_z/p_0)^2 \sim \epsilon_d/E_d \sim 1$ percent. It can readily be verified explicitly that this correction term is negligible.

The probability of a given value of $p_\perp$ is

$$P(p_\perp)2\pi p_\perp dp_\perp = \int_{-\infty}^{\infty} \frac{(M\epsilon_d)^{\frac{1}{2}}dp_z}{\pi^2(M\epsilon_d+p_z^2+p_\perp^2)^2} \cdot 2\pi p_\perp dp_\perp$$

$$= \frac{(M\epsilon_d)^{\frac{1}{2}}}{2\pi(M\epsilon_d+p_\perp^2)^{\frac{3}{2}}} \cdot 2\pi p_\perp dp_\perp.$$

Expressing this in terms of θ by means of (10), we find

$$P(\theta)d\Omega = (1/2\pi)[\theta_0/(\theta_0^2+\theta^2)^{\frac{3}{2}}]d\Omega, \quad (11)$$

where

$$\theta_0 = (\epsilon_d/E_d)^{\frac{1}{2}}[1-(E_d/8Mc^2)], \quad (12)$$

and $d\Omega = 2\pi\theta d\theta$ is the element of solid angle. Or, if angles are measured in terms of θ_0, $\zeta = \theta/\theta_0$,

$$P(\zeta)d\Omega_\zeta = (1/2\pi)[1/(1+\zeta^2)^{\frac{3}{2}}]d\Omega_\zeta, \quad (13)$$

with $\Omega_\zeta = 2\pi\zeta d\zeta$.

A graph of $2\pi P(\zeta)$ is given in Fig. 2. $P(\zeta)$ falls to half its maximum value at $\zeta = 0.7664$. Thus the half-width of the angular distribution (full width at half-maximum) is $\zeta_{\frac{1}{2}} = 2 \times 0.7664 = 1.533$, or $\theta_{\frac{1}{2}} = 1.533\theta_0$.

The energy of the emergent neutron is

$$E = (1/2M)[(p_0+p_z)^2+p_\perp^2]$$
$$= (1/2M)[p_0^2+2p_0p_z+p^2].$$

Since, for the main part of the distribution, $p_0 \gg p$, we may neglect the last term and write

$$E = \tfrac{1}{2}E_d + (E_d/M)^{\frac{1}{2}}p_z. \quad (14)$$

It will be noted that, while the angular distribution depends on $p_\perp$, the energy distribution depends on p_z.

From (9) we find for the probability of a given p_z

$$P(p_z)dp_z = (2/\pi)(M\epsilon_d)^{\frac{1}{2}}\int_0^{\infty} \frac{p_\perp dp_\perp}{(M\epsilon_d+p_z^2+p_\perp^2)^2}dp_z$$

$$= \frac{(M\epsilon_d)^{\frac{1}{2}}}{\pi(M\epsilon_d+p_z^2)}dp_z.$$

Changing variables from p_z to E by means of (14), we find for the energy distribution

$$P(E)dE = \frac{(\epsilon_d E_d)^{\frac{1}{2}}}{\pi[(E-\tfrac{1}{2}E_d)^2+\epsilon_d E_d]}dE. \quad (15)$$

This gives an energy distribution centered around $E = \tfrac{1}{2}E_d$, with a half-width $\Delta E_{\frac{1}{2}} = 2(\epsilon_d E_d)^{\frac{1}{2}} = 41$ Mev for $E_d = 190$ Mev. A plot of Eq. (15) is given by the dotted curve in Fig. 3.

The extreme tails of this distribution are, of course, not to be believed, in particular the part for which $E > E_d$, which violates energy conservation. Here it is no longer true that $p_0 \gg p$, and, concomitantly, our assumption that the collision of the proton with the nucleus can be regarded as sudden, with no reaction on the neutron, is evidently no longer valid.

We now turn to the calculation of the angular distribution for an opaque nucleus. Referring to Fig. 1, we introduce the coordinates z in the direction of the deuteron, x perpendicular to the edge of the nucleus, and y parallel to the edge of the nucleus. In doing the calculation analogous to (8) we now have to impose the condition that the neutron misses and the proton hits the nucleus, i.e., we are to take $\psi_d = 0$ unless $x_n > 0$ and $x_p < 0$. The y and z integrations in (8) are unaltered; after performing them we are left with a wave function which can be written

$$\psi(p_y, p_z, x_n, x_p) = h^{-1}\int_{-\infty}^{\infty} \psi(p_z', p_y, p_z)$$
$$\times \exp[(i/h)p_z'(x_n-x_p)]dp_z', \quad (16)$$

with $\psi(p_z, p_y, p_z)$ given by (9). We next express the wave function in terms of the momentum variables p_z for the neutron, p_{zp} for the proton:

$$\psi(p_y, p_z, p_z, p_{zp})$$

$$= h^{-1}\int_0^{\infty} dx_n \int_{-\infty}^0 dx_p \psi(p_y, p_z, x_n, x_p)$$
$$\times \exp[-(i/h)(p_z x_n + p_{zp}x_p)]$$

$$= -h^{\frac{1}{2}}(2\pi)^{-\frac{1}{2}}\int_{-\infty}^{\infty} \frac{\psi(p_z'p_yp_z)}{(p_z'-p_z)(p_z'+p_{zp})}dp_z'.$$

The poles in the denominator are to be avoided by deforming the contour of integration into the upper half-plane. Since $\psi(p_x', p_y, p_z)$ has one pole in the upper half-plane, this integral can be evaluated in terms of the residue at the pole. We find

$$\psi(p_y, p_z, p_x, p_{xp}) = \frac{\hbar^{\frac{1}{2}}(M\epsilon_d)^{\frac{1}{2}}}{(2\pi)^{\frac{3}{2}}P(p_x - iP)(p_{xp} + iP)},$$

where

$$P = (M\epsilon_d + p_y^2 + p_z^2)^{\frac{1}{2}}.$$

The probability of a given neutron momentum is

$$P_1(\mathbf{p}) = \int_{-\infty}^{\infty} |\psi(p_y, p_z, p_x, p_{xp})|^2 dp_{xp}$$

$$= \frac{\hbar(M\epsilon_d)^{\frac{1}{2}}}{8\pi^3 P^2(p_x^2 + P^2)} \int_{-\infty}^{\infty} \frac{dp_{xp}}{(p_{xp}^2 + P^2)}$$

$$= \frac{\hbar}{8\pi^2} \frac{(M\epsilon_d)^{\frac{1}{2}}}{P^3(p_x^2 + P^2)}. \tag{17}$$

Equation (17) gives the differential cross section per unit length of the circumference of the nucleus. We have now to integrate around the circumference, in analogy with the integral over dl in the derivation of the total cross section. We must remember that the x and y axes rotate as we go around the nucleus. If ϕ is the azimuthal angle around the circumference, we can write $p_x = p_\perp \cos\phi$, $p_y = p_\perp \sin\phi$, $dl = Rd\phi$, and the differential cross section is

$$d\sigma = \frac{\hbar(M\epsilon_d)^{\frac{1}{2}}R}{8\pi^2(M\epsilon_d + p^2)}$$

$$\times \int_0^{2\pi} \frac{d\phi}{(M\epsilon_d + p_z^2 + p_\perp^2 \sin^2\phi)^{\frac{3}{2}}} d\mathbf{p}. \tag{18}$$

To get the angular distribution, we integrate (18) over p_z,

$$d\sigma = [\hbar(M\epsilon_d)^{\frac{1}{2}}R/8\pi^2] \int_0^{2\pi} d\phi \int_{-\infty}^{\infty} \frac{dp_z}{(M\epsilon_d + p_\perp^2 + p_z^2)(M\epsilon_d + p_z^2 + p_\perp^2 \sin^2\phi)^{\frac{3}{2}}} \cdot 2\pi p_\perp dp_\perp. \tag{19}$$

To carry out the integrations,[3] we change variable from p_z to a new variable ψ, defined by

$$\tan\psi = p_z/(M\epsilon_d + p_\perp^2 \sin^2\phi)^{\frac{1}{2}}.$$

The double integral in (19) becomes

$$2\int_0^{\pi/2} \cos^2\psi d\psi \int_0^{2\pi} \frac{d\phi}{(M\epsilon_d + p_\perp^2 \sin^2\phi)(M\epsilon_d + p_\perp^2 \cos^2\psi + p_\perp^2 \sin^2\psi \sin^2\phi)}$$

$$= \frac{2}{(M\epsilon_d + p_\perp^2)} \int_0^{\pi/2} \cos\psi d\psi \int_0^{2\pi} \left[\frac{1}{M\epsilon_d + p_\perp^2 \sin^2\phi} - \frac{\sin^2\psi}{M\epsilon_d + p_\perp^2 \cos^2\psi + p_\perp^2 \sin^2\psi \sin^2\phi} \right] d\phi$$

$$= \frac{4\pi}{(M\epsilon_d + p_\perp^2)^{\frac{1}{2}}} \int_0^{\pi/2} \left[\frac{1}{(M\epsilon_d)^{\frac{1}{2}}} - \frac{\sin^2\psi}{(M\epsilon_d + p_\perp^2 \cos^2\psi)^{\frac{1}{2}}} \right] \cos\psi d\psi$$

$$= \frac{4\pi}{M^2\epsilon_d^2} \frac{1}{(1+\zeta^2)^{\frac{1}{2}}} \{1 - (1/2\zeta^3)[(1+\zeta^2)\tan^{-1}\zeta - \zeta]\},$$

where ζ is the same variable used in (13), $\zeta = p_\perp/(M\epsilon_d)^{\frac{1}{2}} = \theta/\theta_0$. Putting this in (19) gives

$$d\sigma = [RR_d/\pi(1+\zeta^2)^{\frac{1}{2}}]$$

$$\times \{1 - (1/2\zeta^3)[(1+\zeta^2)\tan^{-1}\zeta - \zeta]\} d\Omega_f. \tag{20}$$

It can readily be verified that integration over $d\Omega_f$ again gives (5) for the total cross section.

[3] I am indebted to Dr. Joseph Weinberg for this integration.

Comparing (20) and (13), we see that the angular distribution for the opaque nucleus differs from that for the transparent nucleus by the factor in the curly bracket. The angular distribution given by (20) is also plotted in Fig. 2, and we see the two distributions are not very different. The half-width given by (20) is $\theta_1 = 1.601\theta_0$, only 4 percent wider than that given by (13). The distribution (20) has a higher tail at large angles than (13), an effect that can be interpreted as caused by the additional diffraction of the neutrons around the edge of the nucleus.

The calculation of the energy distribution of the neutrons parallels the derivation of (15). We first integrate (18) over $p_\perp$; this integral is elementary, and gives

$$\int_0^\infty \frac{p_\perp dp_\perp}{(M\epsilon_d + p_s^2 + p_\perp^2)(M\epsilon_d + p_s^2 + p_\perp^2 \sin^2\phi)^{\frac{1}{2}}}$$

$$= \frac{1}{2(M\epsilon_d + p_s^2)^{\frac{1}{2}} \cos^2\phi} \left\{ \frac{1}{\cos\phi} \ln\left(\frac{1+\cos\phi}{1-\cos\phi}\right) - 2 \right\}.$$

The integration over ϕ gives just a numerical factor, which is equal to π^2. Finally, changing variables from p_s to E by means of (14), (18) becomes

$$d\sigma = \frac{1}{4}\pi RR_d \frac{E_d\epsilon_d}{[(E-\frac{1}{2}E_d)^2 + E_d\epsilon_d]^{\frac{1}{4}}} dE. \quad (21)$$

The energy distribution (21) has a half-width $\Delta E_{\frac{1}{4}} = 1.533(E_d\epsilon_d)^{\frac{1}{2}}$, i.e., $\Delta E_{\frac{1}{4}} = 31$ Mev for $E_d = 190$ Mev. A plot of the energy distribution for this

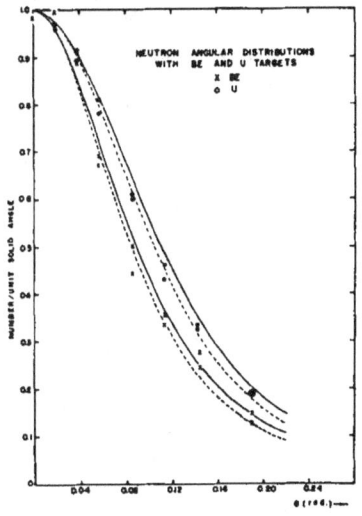

FIG. 4. Measured and calculated angular distributions for Be and U targets. Experimental points from measurements by Helmholz, McMillan, and Sewell. Curves for opaque nucleus (solid line) and transparent nucleus (dotted line) are included for comparison.

deuteron energy is given by the solid curve in Fig. 3. It will be remarked that some neutrons are to be expected with an energy considerably larger than $\frac{1}{2}E_d$: 3 percent of the area of the curve lies between 150 Mev and 190 Mev.

Comparison of (15) and (21) shows that the energy distribution, unlike the angular distribution, is appreciably different in the transparent and opaque cases, the latter giving a narrower distribution.

IV. EFFECT OF THE COULOMB FIELD ON THE ANGULAR DISTRIBUTION

The observed half-widths of the neutron distribution are found to increase slowly with atomic number. This effect can be understood as being caused by the nuclear Coulomb field, which deflects the deuteron slightly before the stripping process takes place. There are two sources of deflection. The first is an intrinsic one: the bending of the deuteron's orbit in the field of the nucleus at whose surface the deuteron is stripped. The second is due to the finite thickness of the target; multiple scattering produces a fanning out of the deuteron beam as it traverses the target. The angle of deflection attributable to the first cause is[4]

$$\theta_c = E_b/2E_d,$$

where $E_b = Ze^2/R$ is the barrier height. For 190-Mev deuterons bombarding U, $\theta_c = 0.037$. The angle of deflection caused by multiple scattering in passing through the target can be calculated from the usual formula[5]; the results for a number of elements are tabulated in the preceding paper by Helmholz, McMillan, and Sewell. Although both angles are small, they are by no means negligible compared to the width of the neutron distribution.

Since the angular distributions given by (20) and (13) are very little different, we shall simplify our treatment of the Coulomb effects by treating the transparent nucleus case. The results can then be translated to the opaque case by multiplying by the factor by which (20) and (13) differ.

---

[4] Since $Ze^2/\hbar v = 1.5$ for 190-Mev deuterons and a U target, it seems adequate to use a classical description to obtain an estimate of the spreading due to the Coulomb field.

[5] E. J. Williams, Proc. Roy. Soc. **169**, 531 (1939).

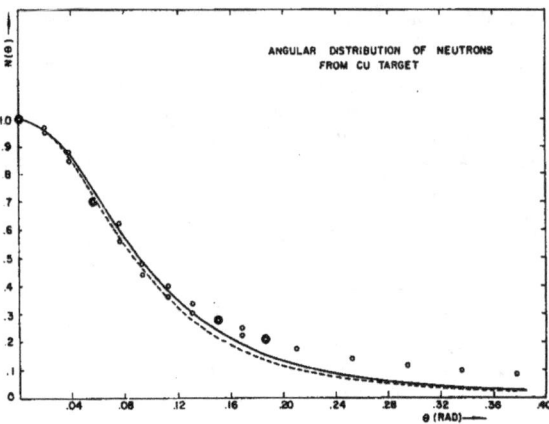

FIG. 5. Measured and calculated angular distributions for a Cu target. Solid curve, opaque nucleus; dotted curve, transparent nucleus. The fact that at large angles the experimental points lie above the curves is presumably due to the production of neutrons, with a wider angular distribution, by processes other than stripping.

We first treat the effect of the deflection by the field of the nucleus which does the stripping. Using the same coordinate system employed in the calculation of the angular distribution from an opaque nucleus, the effect of the bending of the deuteron's orbit is to give the neutron an added momentum $p_0\theta_e$ in the x direction. The distribution function (9) is altered by having p_x replaced by $p_x - p_0\theta_e$. Equation (13) is changed to

$$P(\zeta)d\Omega_\zeta = \int_0^{2\pi} \frac{d\phi}{2\pi[1+\zeta^2\sin^2\theta+(\zeta\cos\theta-\zeta_e)^2]^{\frac{3}{2}}}\zeta d\zeta \quad (22)$$

$$= \int_0^{2\pi} \frac{d\phi}{2\pi[1+\zeta^2+\zeta_e^2-2\zeta\zeta_e\cos\phi]^{\frac{3}{2}}}\zeta d\zeta,$$

where

$$\zeta_e = p_0\theta_e/(M\epsilon_d)^{\frac{1}{2}} = \frac{1}{2}E_b/(E_d\epsilon_d)^{\frac{1}{2}}.$$

The integral appearing in (22) is a Legendre function[*] of argument

$$u = \frac{1+\zeta^2+\zeta_e^2}{[(1+\zeta^2+\zeta_e^2)^2-4\zeta^2\zeta_e^2]^{\frac{1}{2}}},$$

$$P(\zeta)d\Omega_\zeta = \frac{u^{\frac{3}{2}}P_{\frac{1}{2}}(u)}{2\pi(1+\zeta^2+\zeta_e^2)^{\frac{3}{2}}}d\Omega_\zeta. \quad (23)$$

The function $P_{\frac{1}{2}}(u)$ is given by the rapidly

[*] See, for example, E. T. Whittaker and G. N. Watson, *A Course of Modern Analysis* (Cambridge University Press, Teddington, England, 1920), p. 314.

convergent series

$$P_{\frac{1}{2}}(u) = 1 + \frac{3}{4}\left(\frac{u-1}{2}\right) - \frac{15}{64}\left(\frac{u-1}{2}\right)^2$$
$$+ \frac{35}{256}\left(\frac{u-1}{2}\right)^3 - \cdots.$$

For the purpose of making the further correction for multiple scattering, it is more convenient, and sufficiently accurate, to expand the integrand of (22) in powers of ζ_e. This gives

$$P(\zeta)d\Omega_\zeta = [1/2\pi(1+\zeta^2)^{\frac{3}{2}}]$$
$$\times\{1+\zeta_e^2f_1(\zeta)+\zeta_e^4f_2(\zeta)\}d\Omega_\zeta, \quad (24)$$
where

$$f_1(\zeta) = 3(\tfrac{3}{2}\zeta^2-1)/2(1+\zeta^2)^2,$$

$$f_2(\zeta) = 15[1-5\zeta^2+(15/8)\zeta^4]/8(1+\zeta^2)^4.$$

Since (22) is symmetrical in ζ and ζ_e, a better approximation for $\zeta < \zeta_e$ can be obtained simply by interchanging ζ and ζ_e in (24).

The effect of multiple scattering is to give the deuteron beam, after traversing a thickness t of the target, a Gaussian spread of directions with a mean-square angle of scattering proportional to t. The angles θ_s quoted by Helmholz, McMillan, and Sewell are the root mean-square angles after traversing the full target thickness, T. At thickness t, the mean square angle of the Gaussian distribution is thus θ_s^2t/T, or expressed in terms of the spread in ζ, $\zeta_e^2 t/T$, with $\zeta_e = \theta_e(E_d/\epsilon_d)^{\frac{1}{2}}$. The effect of spreading the distribution (24) by this additional Gaussian distribution is given by applying to (24) the integral operator $G = \exp\{(\zeta_e^2 t/4T)\Delta\}$, where

$$\Delta = (\partial/\zeta\partial\zeta)(\zeta\partial/\partial\zeta)$$

is the Laplacian operator. Since the width of the Gaussian distribution is small compared to the width of the distribution (24), the exponential can be expanded, and the operator written

$$G = \left\{1 + \frac{\zeta_e^2 t}{4T}\Delta + \frac{\zeta_e^4 t^2}{32T^2}\Delta^2\right\}.$$

Averaging over the thickness of the target, we get

$$G = \left\{1 + \frac{1}{8}\zeta_e^2\Delta + \frac{1}{96}\zeta_e^4\Delta^2\right\}.$$

Applying this operator to (24), and to the corresponding formula with ζ and ζ_e interchanged, we find

$$
\begin{aligned}
P(\zeta)d\Omega_\zeta &= \frac{1}{2\pi(1+\zeta^2)^{\frac{3}{2}}}\{1 + (\zeta_e^2 + \frac{1}{2}\zeta_e^2)f_1(\zeta) \\
&\quad + (\zeta_e^4 + 2\zeta_e^2\zeta_e^2 + \frac{3}{8}\zeta_e^4)f_2(\zeta)\}d\Omega_\zeta, \quad \zeta \geqq \zeta_e, \\
&= \frac{1}{2\pi(1+\zeta_e^2)^{\frac{3}{2}}}\{1 + \frac{1}{2}\zeta_e^2 f_1(\zeta_e) + \frac{3}{8}\zeta_e^4 f_2(\zeta_e) \\
&\quad + [f_1(\zeta_e) + 2\zeta_e^2 f_2(\zeta_e)]\zeta^2\}d\Omega_\zeta, \quad \zeta < \zeta_e.
\end{aligned}
\tag{25}
$$

Equation (25) gives the neutron angular distribution for the transparent model; the formula for the opaque model is obtained by multiplying (25) by the correction for the opacity of the nucleus, the factor in curly brackets in (20).

The increase in half-width caused by the intrinsic scattering is, in first approximation, proportional to ζ_e^2, i.e., to $E_b^2 \sim Z^2/R^2 \sim Z^2/A^{\frac{2}{3}}$. The increase caused by multiple scattering is proportional to ζ_e^2, or, for given target thickness, to $Z^2\rho/A$. Because of the factor ρ this is not a smooth function of atomic number. With the $\frac{1}{16}$-in. thick targets used in the experiments, the intrinsic Coulomb effect contributes between 90 percent (in Be) and 60 percent (in U) of the total increase in half-width.

The Coulomb field will widen the proton distribution even more than the neutron distribution, since the proton in leaving the nucleus is bent through twice the angle the deuteron is in approaching it. The intrinsic Coulomb effect here is much larger than the multiple scattering, and the angular distribution can be obtained from (23), with ζ_e replaced by $3\zeta_e$.

An additional, smaller effect of the Coulomb field has been pointed out to me by Professor McMillan. The kinetic energy of the deuteron when it reaches the nucleus has been reduced by the amount E_b. Thus E_d in the foregoing formulae is to be taken not as the bombarding energy (corrected for energy loss in the target), but as this minus E_b. An interesting consequence is that the center of the neutron energy distribution will be shifted to lower energy by an amount $\frac{1}{2}E_b$ (7 Mev in U), while the proton, since it regains the energy E_b in escaping, will have its energy distribution shifted upwards by this amount.

Figures 4 and 5 show the calculated neutron angular distributions for targets of Be, U and Cu, and the distributions measured by Helmholz, McMillan, and Sewell. Figure 3 of their paper shows the measured half-widths for a number of elements, and the calculated half-widths The agreement is seen to be quite satisfactory.

A word remains to be said about the neutrons produced by the Coulomb-field disintegration of the deuteron. As previously mentioned, estimates by Dancoff indicate that about one-quarter as many neutrons would be produced in this way as by stripping. Whether, and to what extent, the experimental data might be taken to show the smallness of such an effect depends on the expected angular distribution of the neutrons. This will be the subject of a forthcoming paper by Professor Dancoff, whom I wish to thank for a number of discussions of the electric-field breakup, as well as of the stripping effect.

I am indebted also to Mr. T. B. Taylor, who carried out most of the computations.

This paper is based on work preformed under Contract No W-7405-Eng-48, with the Atomic Energy Commission, in connection with the Radiation Laboratory, University of California, Berkeley, California.

PHYSICAL REVIEW   VOLUME 72, NUMBER 11   DECEMBER 1, 1947

Nuclear Reactions at High Energies

R. SERBER

Radiation Laboratory, Department of Physics, University of California, Berkeley, California

(Received October 13, 1947)

THE general features of the high energy nuclear reactions which have been observed at the Radiation Laboratory can be understood in terms of a picture which is in its main outlines quite simple, though quite different from the description appropriate at lower energies. In trying to understand what happens when a nucleus is bombarded by a high energy neutron or proton, the first consideration that comes to mind is that the collision time between the incident particle and a particle in the nucleus is short compared to the time between collisions of the particles in the nucleus. This suggests that the first step in the process can be regarded in terms of collisions between the incident particle and the individual nuclear particles. We are thus led to ask the properties of the high energy scattering between free nucleons. There are two salient points. First, the total cross section for scattering of one nucleon by another is inversely proportional to the energy of the incident particle. The mean free path of a nucleon traversing nuclear matter increases with its energy; at sufficiently high energies the nucleus begins to be transparent to the bombarding particles. Secondly, the incident particle loses only a small fraction of its energy to the struck one. The momentum transfer, which is nearly perpendicular to the direction of the incident particle, is of the order $\hbar/a$, where a is the range of nuclear forces, and does not increase with increasing energy. In case of an exchange collision, we continue to call the high energy emergent particle the incident one, even though it has changed its charge.

Since the momentum transfer, $\hbar/a$, is not large compared to the characteristic momentum, $\hbar/d$, of particles in the nucleus (with d the mean separation of nuclear particles), it is not true that the collisions made by the incident particle can be considered as collisions between free particles; interference between particles in the nucleus can be important. Such an interference effect can be expected because of the degeneracy of nuclear matter. If nuclear matter is represented as a degenerate Fermi gas, it is clear that collisions with small momentum transfers will be discouraged, since these tend to lead from an occupied state to another already occupied. An estimate of this effect indicates that as a result the mean free path of a high energy particle ($\sim$100 Mev) traversing nuclear matter will be increased over what would be expected for collisions between free particles by a factor of about 5/3. The mean kinetic energy transfer to the struck particle per collision is increased in the same ratio.

We estimate that the mean free path for a 100-Mev nucleon is about 4×10^{-13} cm, and the kinetic energy transfer to the struck particle is about 25 Mev. Since the mean free path is comparable to nuclear radii, one cannot describe what goes on in terms of formation of a compound nucleus. In fact, what happens will depend on the particular trajectory of the incident particle. If it happens to pass through the nucleus near its edge, it may make a single collision and emerge having lost only 25 Mev of its energy, possibly having changed from neutron to proton (or vice versa) as a result of an exchange collision. Or, if it strikes the center of the nucleus and has to pass through the full diameter, it may make several collisions, lose all its energy, and end its range still inside the nucleus. There are thus a variety of possibilities, ranging from the bombarding particle emerging with most of its energy intact to the loss of the entire incident energy to the nucleus.

Since the struck particles have much lower energy and shorter mean free path than the incident one, they can escape from the nucleus without further collisions only if the collision occurs near the edge of the nucleus with the struck particle heading outwards. In this case it may emerge with 15- or 20-Mev energy. Otherwise it will collide with other nuclear particles, the energy will be distributed over the nucleus, and the subsequent events can be described in terms of the usual evaporation model, the nuclear excitation energy being dissipated by successive boiling off of particles each with a few million volts of kinetic energy.

When the nucleus is bombarded with 200-Mev deuterons, or 400-Mev α-particles, the binding of the incident nucleons is important chiefly in causing a spatial correlation between them, and what goes on can be thought of in terms of a simultaneous bombardment by several individual nucleons.

The description we have given provides an explanation of several features of high energy reactions which have been observed at the Radiation Laboratory. Because of the wide distribution of excitation energies of the struck nucleus, one would expect a wide distribution of residual nuclei after the evaporation processes are complete; loss of a small number of particles should occur, as well as knocking out of many. This feature of high energy reactions has been reported by Seaborg, Perlman, and their collaborators.[1] Then we may ask about the excitation function of a particular reaction leading to a given residual nucleus. At low energies the reaction proceeds through formation of a compound nucleus. If we confine ourselves to this mechanism, the excitation function will go through a maximum when the excitation energy is most appropriate for evaporating the requisite number of particles, then will drop very rapidly as the energy gets higher because at higher excitation energy it is much more likely that more particles will evaporate. However, at high energies the mechanism of the reaction is different, because of the transparency of the nucleus; the reaction can occur through the incident particle carrying off a good fraction of its energy and giving the nucleus approximately the right excitation energy for the reaction in question. Since the probability of leaving a given excitation energy will be determined only by the mean free path, which varies slowly with the energy of the incident particle, we would expect the excitation function at high energies also to vary quite slowly. This has been confirmed in a number of cases.[2]

Finally, we should expect the transparency of nuclear matter to show up in measurements of the total cross section for absorption or scattering of the incident particle. It should be mentioned that the attenuation of the wave representing the incident particle in passing through the nucleus will give rise to diffraction scattering at small angles ($\theta \sim \hbar/Rp$, where R is the nuclear radius, p the momentum of the bombarding particle). The cross section for diffraction scattering is equal to the inelastic and absorption cross section; for good geometry attenuation measurements it just doubles the cross section. For 100-Mev neutrons, with a mean free path of 4×10^{-13} cm, one sees that for the heaviest elements one would expect a total cross section still close to $2\pi R^2$, but for light elements the cross section should drop considerably below this value. This is, in fact, true, as has been shown by experiments by Cork, McMillan, Peterson, and Sewell.

A number of more detailed calculations, based on the considerations given above, have been carried out by members of the theoretical group at this laboratory. Reasonably good agreement has been obtained with experimental results on the excitation curves and absolute cross sections of a few light element reactions, and on curves of star size *versus* frequency which have been measured by E. Gardner.[3] More detailed reports on this work will be published in the near future.

This paper is based on work performed under Contract W-7405-Eng-48 with the Atomic Energy Commission in connection with the Radiation Laboratory, University of California, Berkeley, California.

[1] B. B. Cunningham, H. H. Hopkins, M. Lindner, D. R. Miller, P. R. O'Connor, I. Perlman, G. T. Seaborg, and R. C. Thompson, Phys. Rev. **72**, 739 (1947).

[2] W. Chupp and E. M. McMillan; R. Thornton and R. W. Senseman, to be published.
[3] Eugene Gardner, Phys. Rev. **72**, 743 (1947).

Reprinted from The Physical Review, Vol. 75, No. 9, 1352–1355, May 1, 1949

The Scattering of High Energy Neutrons by Nuclei

S. Fernbach, R. Serber, and T. B. Taylor

Radiation Laboratory, Department of Physics, University of California, Berkeley, California

(Received January 19, 1949)

The experiments of Cook, McMillan, Peterson, and Sewell on the cross sections of nuclei for neutrons of about 90 Mev indicate that the nuclei are partially transparent to high energy neutrons. It is shown that the results can be explained quite satisfactorily using a nuclear radius $R = 1.37 A^{\frac{1}{3}}$ $\times 10^{-13}$ cm, a potential energy for the neutron in the nucleus of 31 Mev, and a mean free path for the neutron in nuclear matter of 4.5×10^{-13} cm. This mean free path agrees with that estimated from the high energy n-p cross section, but the results are not sensitive to the choice of mean free path.

IN a previous paper by one of the writers[1] it has been pointed out that to a high energy bombarding particle a nucleus appears partially transparent, since at energies of the order of 100 Mev the scattering mean free path for a neutron or proton traversing nuclear matter becomes comparable to the nuclear radius. This transparency effect is strikingly apparent in the experiments of Cook, McMillan, Peterson, and Sewell[2] on the scattering by nuclei of neutrons of about 90 Mev. In the present paper it will be shown that the observed scattering cross sections can be quite satisfactorily accounted for, using a mean free path of the expected magnitude.

The problem is that of the scattering of the neutron wave by a sphere of material characterized by an absorption coefficient and an index of refraction. The index of refraction is determined by the mean potential energy, V, of the neutron in the nucleus. If $k = (2ME)^{\frac{1}{2}}/h$ is the propagation vector of the wave outside the nucleus, its propagation vector inside is $k + k_1$, with

$$k_1 = k[(1 + V/E)^{\frac{1}{2}} - 1].$$

For $E = 90$ Mev, $k = 2.08 \times 10^{12}$ cm^{-1}. The potential V is generally taken to be about 8 Mev larger than the energy of the Fermi sphere. The latter depends on the assumed nuclear density. If we use for the nuclear radius the value $R = 1.37 A^{\frac{1}{3}} \times 10^{-13}$ cm,

deduced by Cook, McMillan, Peterson, and Sewell from the 14–25 Mev scattering results of Amaldi, Bocciarelli, Cacciapuoti, and Trabacchi,[3] and Sherr,[4] we find a Fermi energy of 22 Mev, and $V = 30$ Mev. This gives $k_1 = 3.22 \times 10^{12}$ cm^{-1}. The absorption coefficient in nuclear matter is equal to the particle density times the cross section for scattering the neutron by a particle in the nucleus,

$$K = 3A\sigma/4\pi R^3.$$

In terms of the n-p and n-n cross sections,

$$\sigma = [Z\sigma_{np} + (A - Z)\sigma_{nn}]/A.$$

Cook *et al.*[2] give for the scattering of a 90 Mev neutron by a free proton $\sigma_{np(\text{free})} = 8.3 \times 10^{-26}$ cm^2. This cross section must be reduced to allow for the effect of the exclusion principle on the scattering by a proton bound in the nucleus; according to Goldberger,[5] the factor is $\sigma_{np} = \frac{2}{3}\sigma_{np(\text{free})}$. Assuming a $1/E$ dependence of the cross sections we find, for $E = 90 + 30 = 120$ Mev, $\sigma_{np} = 4.15 \times 10^{-26}$ cm^2. If, following Goldberger, we take $\sigma_{nn} = \frac{1}{3}\sigma_{np}$, and use the previously quoted radius formula, we obtain $K = 2.4 \times 10^{12}$ cm^{-1} for $Z/A = \frac{1}{2}$, $K = 2.1 \times 10^{12}$ cm^{-1} for $Z/A = 0.39$ (U). It will be seen from these numbers that in the ensuing calculations it will be a reasonable approximation to suppose that $kR \gg 1$, but k_1/k and $K/k \ll 1$, so that $k_1 R$ and KR are of order one.

[1] R. Serber, Phys. Rev. **72**, 1114 (1947).
[2] L. J. Cook, E. M. McMillan, J. M. Peterson, and D. C. Sewell, Phys. Rev. **75**, 7 (1949).
[3] E. Amaldi, D. Bocciarelli, B. N. Cacciapuoti, and G. C. Trabacchi, Nuovo Cimento **3**, 203 (1946).
[4] R. Sherr, Phys. Rev. **68**, 240 (1945).
[5] M. L. Goldberger, Phys. Rev. **74**, 1268 (1948).

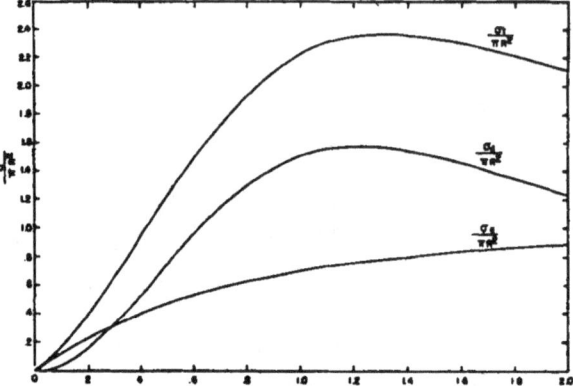

Fig. 1. Absorption, diffraction and total cross section as a function of the nuclear radius measured in mean free paths. These curves are for $k_1/K = 1.5$.

The scattering cross section consists of two parts. The first, the "absorption cross section," is just πR^2 times the probability that the neutron collides with a particle in the nucleus. This is not true absorption: inelastic scattering and scattering with exchange are included. The second part, the "diffraction scattering," is elastic scattering arising from the disturbance of the incident plane wave by the nucleus. To illustrate the calculation, we first consider the scattering from a disk of radius R and thickness T. We suppose there is a boundary layer at the surface of the disk in which k_1 and K rise to their interior values in a distance larger than $1/k$.[6] There will then be no scattering at the surfaces, and, for unit amplitude of incident wave, the wave transmitted through the disk will have an amplitude and relative phase $a = \exp(-\frac{1}{2}K + ik_1)T$. The absorption cross section is

$$\sigma_a = \pi R^2(1 - |a|^2) = \pi R^2(1 - e^{-KT}). \quad (1)$$

The diffraction cross section can be found from the consideration that on a plane behind the disk the wave is no longer plane, but differs from a plane wave by an amplitude $1 - a$ in the shadow of the disk. This amplitude represents a scattered wave, and the corresponding cross section is

$$\sigma_d = \pi R^2 |1 - a|^2$$
$$= \pi R^2(1 - 2e^{-\frac{1}{2}KT} \cos k_1 T + e^{-KT}). \quad (2)$$

It can easily be shown that the angular dependence of the scattered amplitude is

---

[6] In terms of the model being employed, the finite intercept of the R vs. $A^{\frac{1}{3}}$ line obtained from the data on the lower energy scattering could be interpreted by the more careful examination of the boundary conditions which in this case would be necessary.

$$f(\theta) = k \int_0^R (1-a) J_0(k\rho \sin\theta)\rho\, d\rho$$
$$= (1-a)R J_1(kR \sin\theta)/\sin\theta, \quad (3)$$

which gives the differential scattering cross section

$$d\sigma_d(\theta) = |f(\theta)|^2 d\Omega$$
$$= (\sigma_d/\pi)[J_1(kR \sin\theta)/\sin\theta]^2 d\Omega. \quad (4)$$

The absorption cross section is, of course, always less than πR^2, but the diffraction cross section may be either larger or smaller, depending on the magnitude of the phase shift. For large KT, $\sigma_a = \sigma_d = \pi R^2$. In the opposite limit of small KT and k_1T, we have

$$\sigma_a = \pi R^2 KT = A\sigma,$$
$$\sigma_d = \pi R^2(\frac{1}{4}K^2 + k_1^2)T^2 = \frac{1}{4}A\sigma[1 + 4(k_1^2/K^2)]KT.$$

Thus for low density or small thickness, σ_a approaches the sum of the scattering cross sections of the separate nucleons. The diffraction cross section, however, vanishes in the limit, being proportional to the probability of double scattering.

The corresponding calculations for a sphere are only slightly more complicated. The portion of the wave which strikes the sphere at a distance ρ from a line through the center of the sphere emerges after traveling a distance $2s$, with $s^2 = R^2 - \rho^2$. Its amplitude on emerging is $a = \exp(-K + 2ik_1)s$, so that, in place of (1) we have

$$\sigma_a = 2\pi \int_0^R (1 - e^{-2K s})\rho\, d\rho = 2\pi \int_0^R (1 - e^{-2K s})s\, ds$$
$$= \pi R^2\{1 - [1 - (1 + 2KR)e^{-2KR}]/2K^2R^2\}. \quad (5)$$

This formula for the absorption cross section has previously been given by Bethe.[7] Similarly, in

---

[7] H. A. Bethe, Phys. Rev. 57, 1125 (1940).

place of (2), we have

$$\sigma_d = 2\pi \int_0^R |1 - e^{(-K+2ik_1)s}|^2 \rho\, d\rho$$

$$= \pi R^2 [1 + (1/2K^2R^2)\{1 - (1+2KR)e^{-2KR}\}$$
$$- (1/(\tfrac{1}{4}K^2+k_1^2)^2R^2)\{(\tfrac{1}{4}K^2-k_1^2)$$
$$+ e^{-KR}[2k_1R(\tfrac{1}{4}K^2+k_1^2)+k_1K]\sin 2k_1R$$
$$- e^{-KR}[(\tfrac{1}{4}K^2-k_1^2)+KR(\tfrac{1}{4}K^2+k_1^2)]$$
$$\times \cos 2k_1R\}]. \quad (6)$$

In deriving (5) and (6) we have neglected refraction

For purposes of calculation, the integral can be converted to a sum; letting $l+\tfrac{1}{2}=k\rho$ and using the relation $J_0((l+\tfrac{1}{2})\sin\theta)=P_l(\cos\theta)$, valid for large l and small θ, we find

$$f(\theta) = \tfrac{1}{2}k \sum_{l=0}^{l+\tfrac{1}{2}<kR} (2l+1)(1-e^{(-K+2ik_1)s})P_l(\cos\theta), \quad (9)$$

where

$$s_l = [k^2R^2-(l+\tfrac{1}{2})^2]^{\tfrac{1}{2}}/k.$$

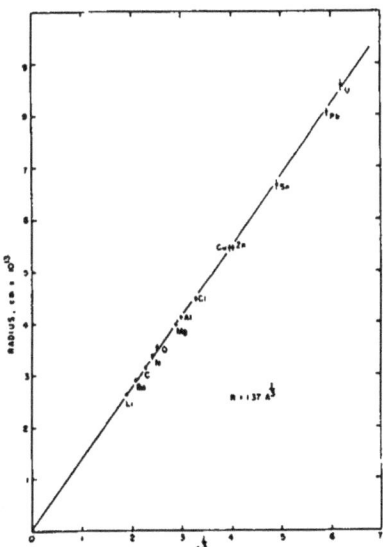

FIG. 2. Nuclear radii deduced from the total cross section measurements of Cook, McMillan, Peterson, and Sewell, plotted against the cube roots of the mass numbers. The straight line is $R=1.37A^{\tfrac{1}{3}}\times10^{-13}$ cm.

at the surface of the sphere. It can easily be seen that this is legitimate, since it gives an effect of order $(k_1/k)k_1R$.

For the angular distribution we find, in analogy to (3),

$$f(\theta) = k \int_0^R [1 - e^{(-K+2ik_1)s}]J_0(k\rho\sin\theta)\rho\, d\rho. \quad (7)$$

For $KR\to\infty$, we again obtain (4), but we have not found a convenient expression in the general case. The amplitude for forward scattering is easily evaluated, and is found to be

$$f(0) = -\frac{kR^2}{2}\left\{1 + \frac{(k_1-\tfrac{1}{2}iK)^2[1-(1+KR-2ik_1R)e^{(-K+2ik_1)R}]}{2(\tfrac{1}{4}K^2+k_1^2)^2R^2}\right\}. \quad (8)$$

This expression can also be obtained by a partial wave analysis, using the WKB method to evaluate the phase shifts. This gives

$$\delta_l = (k_1+\tfrac{1}{2}iK)s_l,$$

whence we immediately obtain (9), and for σ_a and σ_d,

$$\sigma_a = (\pi/k^2)\sum_l (2l+1)(1-e^{-2Ks_l}), \quad (10)$$

$$\sigma_d = (\pi/k^2)\sum_l (2l+1)|1-e^{(-K+2ik_1)s_l}|^2. \quad (11)$$

Converting the sums in (10) and (11) to integrals we again obtain (5) and (6).

In Fig. 1 we have plotted $\sigma_a/\pi R^2$, $\sigma_d/\pi R^2$, and the total cross section $\sigma_t/\pi R^2 = (\sigma_a+\sigma_d)/\pi R^2$ as functions of KR. The ratio $\sigma_a/\pi R^2$ is a function only of KR; the other two depend on k_1/K as well. The curves in Fig. 1 have been plotted for $k_1/K = 1.5$, about the ratio indicated by our earlier consideration of the expected magnitude of the constants. Using this plot it is possible to determine, once a value of K is chosen, the radius required for each nucleus to give the measured total cross section. The radii calculated in this way from the observed cross sections, using the value[8] $K = 2.2 \times 10^{12}$ cm^{-1}, are shown in Fig. 2. It will be seen that they lie quite closely on the line $R = 1.37A^{\tfrac{1}{3}}\times10^{-13}$ cm; the self-consistency of our description of the scattering process is thus established. The value $K = 2.2\times10^{12}$ cm^{-1} corresponds to a mean free path in nuclear matter of 4.5×10^{-13} cm. The associated value, $k_1 = 3.3\times10^{12}$ cm^{-1}, corresponds to $V = 30.8$ Mev.

The question now arises as to the accuracy with which the constants K and k_1 are determined by the scattering data. If k_1 is decreased, keeping K constant, it is found that the radius curve, Fig. 2, is pulled up in the middle; the resultant curve can

[8] The small dependence of K on Z/A is unimportant, as we shall see later.

be approximated by two straight lines, the light elements lying on a steeper line through the origin, while the heavy elements lie on a less steep line with a positive intercept. Increasing k_1 has the opposite effect. A variation in k_1 of $\pm 0.2 \times 10^{12}$ cm^{-1}, or in V of ± 2 Mev, begins to produce appreciable bending. A reduction in K, with fixed k_1, introduces a curvature in the radius line, the center being pulled down and the two ends raised. The curvature becomes noticeable if K is reduced to less than $K = 1.9 \times 10^{12}$ cm^{-1}, however K can be almost doubled before the opposite curvature becomes very pronounced. For example, $K = 3.0 \times 10^{12}$ cm^{-1} gives an about equally good straight line, $R = 1.39 A^{\frac{1}{3}} \times 10^{-13}$ cm. The total cross-section measurements thus determine the potential fairly well, but are quite insensitive to the absorption coefficient. Measurements of σ_a and of the differential diffraction scattering are required for a better evaluation of K. It should be noted that while k_1 and K are determined directly from the cross sections, the evaluation of V depends also on the energy of the incident neutrons. Cook *et al.* state that the energy of the neutrons detected in their experiment may be a little lower than 90 Mev, lying somewhere between 80 and 90 Mev. If we took $E = 80$ Mev we would find $V = 28.8$ Mev.

For $K = 2.2 \times 10^{12}$ cm^{-1}, the values of KR range from 0.58 for Li to 1.87 for U. It will be seen from Fig. 1 that the nuclear opacity, $\sigma_a/\pi R^2$, would vary from 0.52 for Li to 0.88 for U. It will also be seen that over this range of values of KR it would be expected that σ_d will be nearly twice as large as σ_a.

If one plots the angular distribution of the diffraction scattering given by (9) (i.e., $d\sigma_d(\theta)/d\sigma_d(0)$ *versus* $kR \sin\theta$) one finds curves for the heaviest nuclei which are indistinguishable from that for an opaque nucleus (Eq. (4)), at least as far as the first minimum of the diffraction pattern. For the lighter nuclei, the form of the curve is closely the same, but with an altered scale of abscissa, corresponding to using an effective radius somewhat smaller than the true radius. The increase in the half width of the diffraction peak is zero for $KR = 1.78$ (Pb), 3.7 percent for $KR = 1.20$ (Cu), 6.2 percent for $KR = 0.90$ (Al) and 9.6 percent for $KR = 0.63$ (Be). Measurements of the diffraction scattering and of the absorption are now in progress in this laboratory.

Work described in this paper was done under the auspices of the Atomic Energy Commission.

AUTHOR INDEX

SUBJECT INDEX